No Wonder You Wonder!

Claude Phipps
Illustrations by Friedelwolf Wicke

No Wonder You Wonder!

Great Inventions
and
Scientific Mysteries

 Springer

Claude Phipps
Photonic Associates, LLC.
Santa Fe, NM, USA

Illustrations by
Friedelwolf Wicke
Vienna, Australia

ISBN 978-3-319-21679-9   ISBN 978-3-319-21680-5   (eBook)
DOI 10.1007/978-3-319-21680-5

Library of Congress Control Number: 2015947269

Springer Cham Heidelberg New York Dordrecht London
© Springer International Publishing Switzerland 2016
This work is subject to copyright. All rights are reserved by the Publisher, whether the whole or part of the material is concerned, specifically the rights of translation, reprinting, reuse of illustrations, recitation, broadcasting, reproduction on microfilms or in any other physical way, and transmission or information storage and retrieval, electronic adaptation, computer software, or by similar or dissimilar methodology now known or hereafter developed.
The use of general descriptive names, registered names, trademarks, service marks, etc. in this publication does not imply, even in the absence of a specific statement, that such names are exempt from the relevant protective laws and regulations and therefore free for general use.
The publisher, the authors and the editors are safe to assume that the advice and information in this book are believed to be true and accurate at the date of publication. Neither the publisher nor the authors or the editors give a warranty, express or implied, with respect to the material contained herein or for any errors or omissions that may have been made.

Printed on acid-free paper

Springer International Publishing AG Switzerland is part of Springer Science+Business Media (www.springer.com)

This book is dedicated to Dr. Ray Kidder, who taught me that complex science can be expressed simply as he explained inertial confinement fusion with a few equations on the whiteboard, during my first week on the job at Livermore.

Sorèze, May 11, 2015

Acknowledgments

Sincere thanks to Ho Ying Fan, my patient and careful assistant editor at Springer USA who shepherded it through the approval process, and to my editor, Chris Coughlin, and to the Springer organization for entertaining this unusual project among their technical books.

Even more special thanks to my friend and artist Mag. Friedelwolf Wicke, whose work appears throughout *No Wonder You Wonder*, with his special blend of wry humor. This book would have been much more dry and colorless without his work!

Thanks to Mr. Robert Schatzkin of W. W. Norton and Ms. Diana Neuwirth of Universal Uclick for making reasonable deals for the two art works whose use I paid for.

Thanks to those who gave me permission to use their works.

Thanks to my wise companion Shanti Bannwart for comments and encouragement along the way, and for being my first reader, never pulling punches when she didn't like something.

Thanks to Hon. Fred Boettcher, a highschool chum, for his interest in the project as a way to turn kids on to science.

Sincere thanks to the rest of my readers:
Dr. Mick Shaw, formerly Rutherford Appleton Laboratory, United Kingdom
Mr. Joram Wolanow, Dallas, TX, a very careful proofreader
Ms. Mary Sisk, formerly Procter and Gamble, St. Louis, MO
Dr. David Reynolds, Coos Bay, Oregon
Mr. August Railey, Santa Fe, NM
Dr. Ray Kidder, Pleasanton, CA
Mr. Christophe Bonnal, Paris, France
Mr. Jerry Jerome, Phoenix, AZ

Finally, I'd like to thank the lovely village of Sorèze in southern France and Carole Watanabe for welcoming me for a month at the beginning and at the end of the project. It is true that one needs isolation to write, and the first shoots need nourishment.

Contents

Part I Basics

Einstein Said … 5
Do You Care? 6
Moral and Scientific Issues 7

Numbers 9
Mental Blackboard 15

Magic Numbers 17

Introduction: Weird Reality 21
The Big Bang: Dark Matter, Dark Energy, and … Stuff 21
What Is Stuff? 24
Gravity 30
Inertia 34
Relativity at Low Speeds: Kinetic Energy 35
Relativity at High Speeds 36
Time Travel 36
Backward Time Travel 37
Forward Time Travel 39
Teleportation 41
Quantum Mechanics: Weirdness on a Small Scale 43
Electromagnetic Waves 43
Quantum Erasers: True Weirdness 45
Entangled Photons: More Weirdness 47
Uncertainty 47
Superconductivity 50
Zero Point Energy (ZPE) 52
Black Holes 52
4D Ghost Stories 55
Cutting World 56
Conclusion 58

The Metric System	59
Exercise	62
Conclusion	67
Exponentials and Instabilities	69
Instability	71
People Instabilities	72
Conclusions	74

Part II Who *Really* Did It First?

Chinese Science and Art	77
Gunpowder!	77
Printing	79
Navies and Voyages	80
Silk!	82
Conclusion	85
Space Geometry from Ancient Greece to Today	87
Venus Transit: A Side Story that is Not Greek	94
Conclusion	95
Rome	97
Concrete	97
Roman Roads	97
Roman Text Messages	98
"All Roads Lead to Rome"	101
Pont du Gard	101
Conclusion	104
Islamic Science and Art	105
Ibn Sahl	106
Ibn Al Haytham	108
Ibn Musa al Kwarizmi	109
Astrolabe and Astronomy	110
Al-Biruni	111
Al-Razi	111
Omar Khayyam	111
Translating Scientific and Philosophical Works	112
Conclusions	112

Part III Modern Science and Engineering

Modern Science	115
Kepler and Newton	115
Einstein	122
Special Relativity	123

General Relativity: The Final Weirdness	127
Conclusion	128

Why Is the Day Sky Blue? Why Is the Night Sky Black? ... 131
J.C. Maxwell	134
Dark Night Skies	137
Aurorae	138
Conclusion	139

Machines ... 141
Force Multipliers	141
External Combustion Engines	142
Industrial Revolution	145
Carnot and Power Generation	145
Improvements	147
Stirling Engine	148
Steam Turbines	149
Internal Combustion Engine Zoo	150
Otto Cycle	150
Wankel Cycle	150
Radial Engine	151
Diesel	151
Hybrid	152
Tesla	152
Turbine Engines for Cars	153
The Differential	153
Windmills and Renewable Energy	155
Inverse Machines, the Servel and Liquid Helium	155
Electric and Laser Refrigerators	156
The Self-Replicating Machine	156
Other Machines	157
Conclusions	158

Lighter Than Air Flight ... 159
Hot Air Balloons	159
Gas-Filled Balloons	161
Dirigibles	162
Helium, in Texas?	163
High-Altitude Helium Balloons	163
Distance Records	164
Spying	164
Conclusions	164

Electricity ... 165
Volts, Ohms, Amps, and Watts	165
Voltage: Hazardous?	166
Resistance	167

Power	167
Transformers	167
Tesla and the First Wireless	169
Birds on a Wire	171
Different Hazard If It's High Frequency	172
AC and DC	172
Transformers	172
Experiment	173
How Do You Use Electricity?	174
DC Motors	174
AC/DC Motors	174
AC Motors	175
Induction Motor	176
How Do You Make Electricity?	176
Batteries	176
Solar Cells	177
Generators	177
Faraday's Homopolar Machine	178
Experiment	178
Electromagnetic Tethers	180
The Earth Dynamo	181
Lightning!	181
Experiment: Liquid Van de Graaf	182
Sprites	183
Wimshurst Machines and Capacitors	184
Conclusion	185
Drones and Robotic Flight	187
Robotic Warriors	188
Conclusions	189
The Jet Plane: How Metal Birds Fly	191
First Commercial Flight: The Comet	192
How Can a Metal Bird Fly?	194
Wings	195
The Turbojet	197
Supersonic Commercial Jet Planes	203
Other Supersonic Jets	205
Pulse Jets	207
Even More Advanced Jet Engines: The Ramjet	208
Conclusions	209
Rockets	211
Rockets vs. Jets	211
Reaction	214
Satellites	216

Sputnik	216
From the Moon to Mars and Far Beyond	220
Exotic Rockets	221
Ion Engines	221
Laser Space Propulsion	222
Photon Rockets	222
Thermonuclear Rockets	222
Satellites: What Holds Them Up?	223
Zero Gravity Environments and Free Fall	223
Zooming Along Without Feeling a Thing	225
Space Junk: The Downside of Satellite Launches	226
What One Invention Will Survive Us?	227
Conclusion	228
Electromagnetic Waves	**229**
Michelson–Morley Experiment	232
Vectors and Fields	234
Maxwell's Equations	235
Radio	236
Transmitters	239
Electric Dipoles	239
Magnetic Dipoles	239
Microwave Flutes	241
Conclusion	241
Electronics and Computers	**243**
Introduction	243
Amplifiers	243
Logic Gates: The Guts of a Computer	245
The First Electronic Computer	246
Computing Speed and Parallel Processing	247
Information Explosion and AI	248
Social Media	250
The Universal Brain	251
iDevices	252
Robots and Drones	252
"Artificial" Beings and Neural Nets	253
Conclusion	255
Biology	**257**
Mendel	257
Darwin	258
"Sequencing" Human DNA	258
Snipping Genes	259
Reassembly: Modern GMO vs. What We've Done All Along	260
Rewriting DNA Code	262
Conclusion	263

Optics 265
Mirrors: The First Selfie 265
Signaling in the Raj 265
Curved Mirrors 267
Lenses: The First Optics 267
Cameras 268
Black and White Photos and Movies 268
Chester Carlson 269
Kodachrome! 270
Polaroid 271
3D 271
Prisms and Spectrometers 273
Diffraction Gratings 275
Spectrometer 275
Telescopes 276
Visible and Infrared Telescopes 276
X-ray Telescopes 277
Gamma Ray Telescopes 278
Microwave and Radio Telescopes 278
Microscopes 278
   Electron Microscopes 278
Fiber Optics 279
Conclusion 281

Lasers 283
Theory 283
Spectra 283
Absorption 284
Quanta 285
Chemistry 285
Emission 286
The Laser Idea 288
Inversion 289
Start of the Laser Age 289
The Laser Zoo 289
What's Special About Laser Light? 291
Ultrashort Pulses 293
Digression: Bats and Chirped-Pulse Acoustic Radar 294
Petawatt Pulses, Attosecond Lasers 295
Applications 296
   Photodynamic Therapy (PDT) 296
   Laser Engraving, Welding and Cutting 297
   3D Metal Printing 298
   Cornea Surgery 298
   Laser Gyros 298

Laser Produced Plasma	300
Orbital Debris Removal	301
Laser Space Propulsion	301
Pyramids of Cheops	301
NIF	302
Laser Fusion Dream	302
Making Integrated Circuits	304
Moore's Law?	304
Modern Fabs	307
LIGO	307
Conclusions	308
Nucleonics	309
Fission	310
Fusion	312
Safety and Utility	312
Fusion Zoo	313
Inertial Fusion	313
Magnetic Fusion	313
Cold Fusion	314
Muon-Catalyzed Fusion	315
Ion Beam Fusion	315
Electrostatic Fusion	315
Conclusion	315

Part IV Odds and Ends

Religion	319
Priesthoods	321
Conclusion	321
Everything Else	323
The Death Asteroid: The Really Big Space Debris Story	323
End of the Dinosaurs	325
Solutions	326
Where Did Water Come From?	327
Finding Exoplanets	328
Planets and Stars	328
What Is Special About Earth?	328
Stars	329
Planets and Exoplanets	329
The Transit Method	330
Radial Velocity	330
Timing	330
Direct Imaging	331
Microlensing	331
Conclusion	331

In Praise of *No Wonder You Wonder!*

"From beginning to end, and with laugh after laugh, I enjoyed every single word of this remarkable book. Phipps is a hell of a good writer, and the kind of physics teacher that I would have loved as a young student. *No Wonder You Wonder* can be engrossing for anyone with a bit of curiosity, not just the scientific minded."
– Christophe Bonnal, Chief Engineer, CNES (French Space Agency)

"*No Wonder You Wonder* is a fantastic book. Covering topics such as space, matter, and the energy within the universe, this book does an excellent job of clarifying these topics. It's a great read for young scientists and aspiring physicists." – August R., high school freshman

"*No Wonder You Wonder* is an utter joy to read! Phipps is able to delve deep into the depths of black holes, the algorithms of time, or the construction of a jet engine while carrying you along on an informative and exciting ride… a 'must read' for anyone interested in a deeper understanding of the world around us." – Jonah Cohen, Sound Engineer, Musician, Entrepreneur

"This book is like a conversation with a favorite uncle who opens doors to secret places we have wondered about, or never knew were out there waiting for us. Phipps is a guide who clearly loves mystery and discovery. He democratizes science by inviting readers to become citizen scientists with agency, awe, and a sense of responsibility." – Garry Hesser, Martin Olav Sabo Professor of Citizenship and Democracy Emeritus, Augsburg College, Minneapolis, MN

"Claude Phipps takes us on a whirlwind tour of the science and technology behind the perennial question, "How does that work?" We learn how science has shaped society from ancient China to the modern day. Phipps emphasizes understanding and working with numbers as a central theme, but in a delightfully conversational tone that teaches and entertains. This whimsically illustrated book is written by one of science's most creative minds whose joy in wondering about the world around him shines through every page." – Mick Shaw, Formerly Director, Krypton Fluoride Laser Programme, Rutherford Appleton Laboratory, UK

"Claude Phipps in No Wonder You Wonder is a modern Plinius the elder summarizing human knowledge up to his time. His book examines math and science in an informative and entertaining way to both those who love math and science and those who don't. You don't know what you didn't know until you've read this book." – Mary Sisk, Retired Procter & Gamble Manufacturing Engineer

"Claude Phipps has written a book that makes science learning fun. He is fascinated and excited by the subject, and it shows in his writing. While exploring a variety of science subjects, he inserts his colorful opinions and anecdotes while properly labeling them as such. He is very careful to avoid burdening the reader with a lot of difficult material, but the reader cannot just breeze through the contents like a novel. It takes time to go over the science and check out the underlying explanations. I recommend this book both for youth and for adults. It is an eye-opener and a mind-opener." – David K. Reynolds, Ph.D., Constructive Living

"I read this book with delight and was fascinated by Phipps' approach! No Wonder You Wonder will be captivating for teachers and their students - and for anyone with a sense of wonder." – Ray Kidder, Laser Research Program Leader, Lawrence Livermore National Laboratory (Ret.)

"Dr. Phipps has written a book as entertaining as it is enlightening. His brilliant scientific mind and playful writing style come together in a fun and fascinating read that will provide endless "ah-ha" moments for anyone who has ever wondered about the many mysteries of our world." – Jerry Jerome, an educator

Part I

Basics

Preface

Lots of scientists can write equations. I can, too! But, as a way of communicating, they are off-putting to anyone who didn't go flying through Algebra II and Calculus, let alone Differential Equations. My goal here is to make the complex simple, instead of the other way round. There is *nothing* you have to read first in this book. You may want to glance at the *Numbers* and *Metric System* chapters first, if you're not comfortable with those things. But you can start anywhere that interests you and read backwards and forwards. That's my kind of book!

Who Am I?

I'm a "retired" scientist in Santa Fe, New Mexico. My expertise is in the physics of pulsed lasers interacting with materials. I have a Ph.D. in plasmas from Stanford, and a Master's from MIT. I worked at the Livermore and Los Alamos labs for many years. So far as I know, I never worked on anything that would hurt someone. I have run a conference called High Power Laser Ablation in Santa Fe every two years since 1998. I still work 50 h a week. Even in science, my intention has always been to make the complex simple. I'm also a poet and writer, and history is one of my favorite things.

When I look out my window at the world, I see an infinitely complex place full of interrelated things. Poke it anywhere, ask a question like "why is the sky blue?," and that question will bore down and branch out into a thousand more questions which are all part of physics. That's why it's such a great field. I don't want to be a narrow expert in anything, but rather knowledgeable about it all. *It is still possible*! That's one reason I wrote this book. I know there are a lot of you out there, who are curious but are not encouraged by the way science is usually taught. You will notice it's written here in a personal, narrative way, and that the same topic shows up in various ways in several chapters, rather than being organized in some hierarchy. For that reason, I've put in links connecting discussions of the same thing. You *can* understand it without getting a degree in it.

The second reason for the book will become more clear in the very first chapter, where I insist that you're not entitled to your own facts. The Internet is a tremendous resource, but it has no filtering mechanism and it's easy for one person's particular opinion to become what "Einstein said."

What is Science and What is Not?

Here's a bit of philosophy: What is science anyway? Is it Francis Bacon's "scientific method," which you may have learned in school: "First, form a hypothesis…"? Of course not! Those ideas were written by Lit and History Majors! Few scientists do science that way. Scientists are human. Science starts in the belly with a hunch, a wish to leave a mark on the world, a sudden crazy thought in a conference, a vision of the benzene molecule in a dream. After I make my theory and think I've proved it, does a contrary result cause me to abandon it? Of course not! All this work was worthwhile after all! If others try to prove me wrong, I'll fight like hell until I *have* to admit they're right!

But, the results of science are not just belief or faith. They're not arbitrary. Science is a reality-based community. Do you follow me? Any good scientific result must tie into reality.

By "reality," I mean this: if there is an agent here *that I can identify and measure, and it acts in the same measurable way everywhere for everyone*, then it's a scientific fact, part of reality. Whether I as a scientist understand it, or not, doesn't matter. It's a gift if I don't: a new relationship to figure out and then tie into the rest of them.

The squishy things happen when I try to explain a scientific fact with a theory about why it happened, because that depends on my imagination and training, and on the state of science in my time. But the foundation of facts is not so squishy. A good scientist is ready to admit that a new and better explanation of things than she learned is possible. After a thousand years of pretty good science, we start from an agreed set of basic ideas ("force is mass times acceleration") that have been proven so often we don't give them a second thought. Why is that? Because experimental scientists have proved these basic theories so many times, in so many different ways, in the lab, some of them out to dozens of decimal places. My theory is no longer squishy when it's proven.

I don't have a right to make up my own facts. Others have to agree.

Science is a continuing process of finding where basic ideas break down and discovering new and better ideas to explain what you can see. For example: you can weigh something and find its mass, and the tick of a clock is the tick of a clock. Yet, as Einstein realized just a century ago, while we can all agree on what we mean by "mass," "speed," "lifetime," and "length" in our own backyards, we will not agree even approximately when I'm going at 90 % of the speed of light away from you! This does not mean, as you may have heard, that "Einstein said everything is relative."

He didn't. His theories of relativity have been proven to microseconds for clocks on satellites circling the Earth, and for multiply extended lifetimes of fast atomic particles that ought not to live as long as they do.

The thrilling part of science is discovering brand new things, like the Higgs boson, which Peter Higgs actually lived to see, or in doing brand new things like landing on the Moon.

It's been a pretty rapid change. Three hundred years ago, you couldn't have bought a battery (although those clever Arabs may have made a few for electroplating 2000 years ago!), and electric lights of any sort were not known until about 1800. Two centuries ago, people were just finding out about electric currents and magnetic and electric fields and how they can produce each other. Electric street and house lights required generators, and these were not perfected until the time of Lincoln. Nobody knew how far away the stars really were until a hundred years ago. Now, the set of scientific facts you will learn in school changes dramatically in one human lifetime. And that is why school takes a long time.

Science is a strong belief system for me. My "belief" is that science is the best way to approach reality.

"Belief" is a funny word. It all comes down to what I think is true. Miracles and magical outcomes are strong belief systems for many, many people. Garcia Marquez got a Nobel prize for stories about that kind of reality and I find his stories beautiful, but that world is a different one from the scientific one.

My observation is that 30–50 % of people believe things that have no scientific basis, and call it "science." If it works for them it would be cruel as well as hopeless to try to disprove their beliefs. Lord knows, at least that fraction of the U.S. population believes the world was made 6000 years ago. Nikita Khrushchev once said, "If the people believe there's a river over there, don't tell them there's no river. Make an imaginary bridge over the imaginary river!" And that is good advice for all of us.

I cannot measure "energy," as that word is used colloquially. People see and feel auras or they don't, and that's anecdotal, not reproducible. Energy, in my field, is measured in joules. The energy of a blue photon is always the same. When those guys at CERN fired two proton beams at each other and found the Higgs boson resonance at 125 billion volts of energy, after millions and millions of shots, to me, it was justified and even a holy event to play the "Ode to Joy" from Beethoven's ninth in the movie "Particle Fever." If you haven't seen that movie, go do it!

There's a deeper fear for me, and that's what this preface is about: when it appears the majority of people in general (not you, dear reader!) are starting to doubt that there is a factual reality and prefer the reality on their favorite blog or hearsay from a friend—"of course, we know that…", "Einstein said that…" (poor Einstein, he said so *many* things)—the entire Enlightenment Experiment, the basis of Modern Civilization is doomed. Yes, I know I already ranted about that in Chap. 1!

I also "believe in" the power of prayer. I cannot measure it, it is certainly anecdotal, and what it is that I am praying to, or how it acts, is a mystery. This is in that other world, that I *believe in but cannot measure*, and *have no need to measure*. It's not true that I don't believe in anything I cannot measure. I believe in love, in beauty, in honesty and so on. I'm sure you do too! But, I also believe in science and repeatable scientific results.

I admit there are many things in the scientific world that nobody understands. And that makes it so interesting. There's *so much* left for your generation to find out! But there *are* things we do understand.

That's why I'm writing this book for you, to help you keep these things separate! It's not just a matter of belief, what is science and what is not. The world of belief and the world of science do *not* contradict each other, although religions have often claimed that they did.

So: onward and upward! Enjoy!

Einstein Said …

And now, as the Monty Python folks used to say, "for something completely different!" Poor Einstein! He said so many things! Many of these are examples of how people like to buttress their own opinions, or, which is better, enhance a story. Glenn Hodges, on his sounding line blog puts it succinctly: "Einstein is the victim of interminable New Agey quote chains that seem to revel in the notion that the greatest scientific mind of the twentieth century believed the same things Oprah does." This is why I'm writing this chapter.

1. "Everything is Relative." He didn't say that. As we say in *Modern Science*, when you come back from a very fast trip, some things—like the age of your friends—are irreversibly changed.
2. One day, someone saw a horseshoe hanging above his door and asked him if he really believed this brought him luck. "I don't believe in it! But it works anyway!" he is supposed to have said.
   Nope. The truth is: Niels Bohr was visiting a friend, saw the horseshoe and asked "Do you really believe in this?" to which his friend replied "Oh, I don't believe in it. But I am told it works even if you don't believe in it." The truth is often *much* less dramatic.
3. "Genius is ten percent inspiration, ninety percent perspiration." Actually, it was Edison, and what he said was "Genius is one percent inspiration, ninety nine percent perspiration," a little bit more boring. But then, he may have put out too much indiscriminate effort in developing light bulbs—scientists sometimes ridicule undiscriminating industry by saying someone's efforts are "Edisonian."
4. "If you can't explain your physics to a barmaid it is probably not very good physics." That's a good one, and I believe it. But it was Ernest Rutherford who said it.
5. "The definition of insanity is doing the same thing over and over and expecting different results." We don't know who said it, but it's probably very old. And it's not a definition of insanity. Maybe of stupidity.

6. "I refuse to believe God plays dice with the Universe." Well, his statement to a colleague was more complex ... "It seems hard to sneak a look at God's cards. But that he plays dice and uses telepathic methods is something I cannot believe for a single moment."
7. "Raffiniert i'st unser Herrgott—aber boshaft ist er nicht." He did say that. It means, "God is particular, but he's not evil."
8. A reporter asked Einstein, "what is the most important question facing humanity today?" To which he replied, "I think the most important question facing humanity is, 'Is the universe a friendly place?' ... For if we decide that the universe is an unfriendly place, then we will use our technology, our scientific discoveries and our natural resources to achieve safety and power by creating bigger walls to keep out the unfriendliness and bigger weapons to destroy all that which is unfriendly, and I believe that we are getting to a place where technology is powerful enough that we may either completely isolate or destroy ourselves as well in this process. But if we decide that the universe is a friendly place, then we will use our technology, our scientific discoveries and our natural resources to create tools and models for understanding that universe. Because power and safety will come through understanding its workings and its motives." That's beautiful, no?

All my friends know this quote. I have found at least 100 versions of it, different in details but basically all quoting each other, without a single source. To me, that's suspicious!

Glenn Hodges continues, "It is impossible to locate a primary source for this quote or any of its several variations, but that doesn't stop a few undistinguished books from offering a version so ballsy that it creates an appearance of authenticity."

As a separate matter, the statement is well-written and worth keeping in mind when we wonder why we've not been contacted by alien beings. And if you find the source, *tell me*!

Do You Care?

Now, here's my question, and I have already asked enough people personally to know that the answer is not clear: *Do You Care* whether Einstein said it or not?

I care very strongly. As I said in the *Preface*, a big part of my life has been devoted to at least knowing what the facts are and where to find them. The whole Enlightenment Experiment, from being able to say the Earth goes around the Sun to making your iPhone, depends on the ability to build knowledge without fear. An electronics engineer must know what is known, and what is not. Otherwise we don't get to the Moon. If you build on BS, that circuit will not work.

To build knowledge, there must be a trusted "library," whether physical or digital, that reliably stores, and permits us to retrieve, the results of the last few hundred years of discovery. In order to do that, this information must be "vetted," as in the laborious peer review process for scientific journal articles. We have to know that person didn't make up her own facts.

When we're entitled to our own facts, we start thinking they staged the moon landing some place in Arizona and that aliens are buried at the bomb test site in Nevada. And that global climate change is a hoax.

The Internet is a tremendous resource and I depend on it—for research, for publication, for communication. But yesterday, when I came across the Einstein quote problem, I suddenly realized what others have, too: it's a giant echo chamber as well. One guy wrote that wonderful paragraph, which stands by itself. However, he felt sufficiently insecure (or greedy—I hope not) to make it Einstein's. It is now accepted knowledge that Einstein wrote that. And it subtly changes what we know about Einstein and his philosophy.

That made me mad, and what was going to be a paragraph became a chapter.

Moral and Scientific Issues

There are both moral and scientific issues here. The moral one is that some of these quotes completely distort what the Swiss scientist would or could have said, changing his history without his consent. And, if you put your words in someone else's mouth, it's a lie.

The scientific one worries me more. Those of us who think quote #8 is beautiful can sit back and feel privileged to live in such a grand world where a great scientist agrees with us. In exactly the same way, those of us who *simply know* that global climate change is a hoax perpetrated by people like me have their own echo chamber, and it's not just limited to flamboyant radio personalities. If you look up "global warming hoax," you'll find all the support you need for that idea. And that's because *people don't care* whether it's true or not, as long as it agrees with them.

You should care. Poetry is one thing. Fake facts are another, very dangerous thing.

I want you to know that *every single thing I've put in this book has a trusted original source that I can defend*, and I want you to realize that you should, and can, look things up rather than just quote them. Or, if they seem fishy, then doubt them, all by yourself.

Just write me at crphipps@photonicassociates.com, and I'll be happy to give you the source. In writing this book, I decided tons of references on every page would get in the road of communicating, just like all those footnotes in most Shakespeare plays.

But I do have them, for each chapter.

Numbers

You do not have to read this chapter! But … if you get in trouble with numbers later, you can come back and get what you need out of it at any time. I put it toward the front so it would be easy for you to find.

You don't need to be embarrassed if numbers have always seemed difficult to you. That's true for almost everybody. And yet, numbers are easier than you think! Big ones, small ones—I'll show you.

Recently, I read about a case in which a hamburger company spent a lot of money advertising 1/3 pound burgers for only slightly more money than the well-known source of 1/4-pounders was charging, only to find in a focus group that people thought they were getting cheated with 1/3 pound. After all, three is less than four, right?!

Does that seem funny to you?

You know that there are three thirds of a pound in a whole pound (a bit over 5 ounces each piece), and four quarters of a pound (4 ounces each), so I hope you know that 1/4 is less than 1/3. Still … equations and big numbers freeze your mind, right? I hope we'll fix that!

I promise never to subject you to equations other than Einstein's famous $E=mc^2$ in this book. But I want you to understand what that equation means, and be able to do calculations with it, from the mass (m) of something and the speed of light (c). We'll talk about big and small numbers, like femtoseconds and Petaflops, and I need you to be comfortable with those, because today's world works with numbers. You can't afford not to know them, and using them is fun. This chapter is about that.

If all of this seems pretty elementary to you, just go on to the next chapter.

Pretty early, we learn about basic numbers: 1, 2, 3, … 10, 100 and, later, 0.1, 0.01, and so on. Did you know that *zero* wasn't even a number until a bit more than 1,000 years ago? Indian mathematicians first thought of zero as a placeholder (for writing numbers like 1000) and Islamic mathematicians like al-Kwarizmi (see the chapter on *Islamic Science*) went on to develop the whole number system we and the Chinese and everyone else use today. This is during the period Europeans call the Dark Ages. We call them Arabic numerals.

Zero means "nothing," which is why people had a hard time understanding it a thousand years ago. How can anything that really exists not exist? Well, it's not really a mind twisting koan after all, just an example of how people twist their own minds. Truth be told, Indian mathematicians had a lot to do with developing those numbers, so really they are Hindu-Arabic, even though you might not recognize them, the way they looked in the 800s AD. Things go back.

Numbers like −1 and −9 are harder to comprehend, because they're less than nothing, but we use them all the time when we want to subtract in our checkbooks to find our account balance.

Also, you know that there are other kinds of numbers that aren't whole, fractions like the weight of those hamburger patties, 1/3 and 1/4. And, when you write them in base 10 (I'll explain later), 1/3 goes on forever: 0.33333333 …, but not ¼. One quarter is just 0.25 and that's it.

The "square root" of two, 1.414212356 … is a fractional number that also goes on without end, just like 1/3, but the digits are all random. Funny as it sounds, the "square root" of a number is the answer to the question, "what number taken times itself is equal to a number?" Taking something times itself is squaring it, so this is called the square root, because roots are the answers to queries like that.

What about π? Pi is just the distance around any circle divided by its diameter, whether the circle is an atomic orbit, a crop circle, or the Earth's Equator. It's also the area of a sphere divided by its diameter squared. Think about that! Why should it be so simple, that one number can describe both things?

Party knowledge: did you know that Pi times ten million is about the number of seconds in a year, to within 1/3 of a percent? "Pi" is equal to 3.14159265 …, another of those endless numbers. Some people with good memories and a need to impress can recite it out to 100 places. Computer nuts have calculated it out to *ten trillion digits* now! It is not a whole number of anything, no matter how far you look, or what base you count in. But it's very useful, and a part of nature! Not to say a "natural number," which is a word people use to talk about the whole numbers like 1, 2, and 3.

Did I say "base" in that last paragraph? The one you know about is base 10, where the digits go from 0 to 9. We use that unless we use base 2. You hackers out there are already comfortable with base 2, where the digits are either 0 or 1. That's because electronic things today are (mostly, still!) either on or off, 0 or 1, and we developed a counting system to match them.

For the rest of us: how does base 2 work? Take a look at Table 1.

If you've never seen "binary" before, can you figure it out? It isn't hard! It's a game, a secret code and, fundamentally, just another bookkeeping system. "Base" is just the thing you're taking powers of to write down big numbers in an efficient way. In binary, that base is 2. Eight is 2^3, so there's a 1 followed by three zeros, just like a thousand in base ten is 10^3, and you write *that* as a 1 followed by three zeros. That is all there is to it. There are little games you use to add and multiply in both systems. You know how to do that in base 10, but I won't get into it here for base 2.

Table 1 Counting in two systems

Decimal	Binary
0	0
1	1
2	10
3	11
4	100
5	101
6	110
7	111
8	1000
9	1001
10	1010
11	1011
12	1100
13	1101
14	1110
15	1111
16	10000

Woops—I'm getting ahead of myself again. What did I mean with that superscript 3? 10^3 just means $10 \times 10 \times 10$, ten times itself three times. The superscript counts the zeros. Just another bookkeeping system. It's easier to write it that way, no?

By writing "c" for the speed of light, we're just using a shorthand code to avoid writing 299,792.458 kilometers/second each time we talk about the speed of light. So "c" is about 300 million meters per second, or 3×10^8. In this chapter, forget about being precise for a while.

Aha! So by c^2, we mean the number which is c times c. That is about nine times 10^{16}. Is that hard to understand? No. Big numbers exist. Instead of writing 10,000,000,000,000,000, we write 10^{16}. Then, you don't have to count 'em. Easier, no? The action happens in that superscript, called an exponent.

Those of you that use Excel know that 10^{16} can also be written 1E16, and writing it that way is better for two reasons: it's even easier when the important number is easier to see for people that use glasses. The "E" in this case is not energy, but just a placeholder for "10 to the ..." My friends who are theoreticians always seem to make graphs with numbers along the axis in 9-point type, so the exponent is 6-point, and anyone who is not sitting in the front row at a conference can't read *the only important part*. That's why I always write 10^{16} as 1E16 in 14-point bold type when I present an Excel chart.

So $E = mc^2$ just means that energy E (in this case) is m (mass) times c times c. If m is 1 kg, the energy E is 9E16 J. A trillion is 1E12, so 9E16 joules is 90 thousand trillion joules! Scientists at Los Alamos decided that the energy of a ton of exploding TNT was 4.18E9 J, or 4.18 GJ. *That's no worse to think about than 4 GHz in*

Fig. 1 Twenty megatons from your coffee cup (DoE public domain)

your computer, right? Big, important numbers. Just 4E9. And Petaflops, which you hear about all the time in the news. To conclude this bit, the energy put out by converting a kilogram of mass into pure energy, 9E16 J, is a bit more than 21 million tons of TNT, 21 MT (Fig. 1). *Now*, you can get a sense for that amount of energy!

To multiply big numbers, just add the exponents! 100 times 100 is 1E2×1E2, and that's 1E4 or ten thousand. You know that. A billion (1E9) is a thousand million (1E3×1E6). In the sixth grade, I wasted whole afternoons trying to multiply big numbers using arithmetic, on paper (we did that back then!). You can do it, but it takes a long time to get the answer to the area of the Earth (Pi times d^2 is 3.14159265 times 12756.328 kilometers times 12756.328 kilometers), and that's a lot more accuracy then you need. Instead, you just multiply the first few numbers out in front and add the exponents. The Earth's diameter is 6.38E3 km within 1 % accuracy, which is good enough for most purposes. By the way, each time you multiply two numbers that are inaccurate, the resulting inaccuracy is a little larger, so the area of the Earth is $3.14 \times (1.27)^2 \times 1E8$, or 5.11E8 km^2 to within about 2 %. That's 500 million square kilometers. Or, 5.11E14 square meters, because there are a million of those in each km^2. By the way, there are ten thousand (1E4) square meters in a hectare, so the Earth has a bit over 50 billion hectares on it, right? And, only a fraction of those hectares can be plowed.

Now, look at Table 2. We also have names for little numbers as well as big ones. Not so hard after all, huh? *Just another bookkeeping system.* Don't ask *me* why the popular names for these go the way they do. They seem to count groups of three

Numbers

Table 2 Abbreviations for powers of ten

Prefix	Prefix	Power of ten (± number of zeros)	Short-hand	Popular number names
-----	---	33	1E33	Decillion
Watta	W	30	1E30	Nondecillion
Xenna	X	27	1E27	Octillion
Yotta	Y	24	1E24	Septillion
Zetta	Z	21	1E21	Sextillion
Exa	E	18	1E18	Quintillion
Peta	P	15	1E15	Quadrillion
Tera	T	12	1E12	Trillion
Giga	G	9	1E9	Billion
Mega	M	6	1E6	Million
Kilo	k	3	1E3	Thousand
Centi	c	−2	1E-2	A hundredth
Milli	m	−3	1E-3	A thousandth
Micro	μ	−6	1E-6	A millionth
Nano	n	−9	1E-9	A billionth
Pico	p	−12	1E-12	A trillionth
Femto	f	−15	1E-15	A millionth of a billionth
Atto	a	−18	1E-18	A billionth of a billionth
Zepto	z	−21	1E-21	A billionth of a trillionth
Yocto	Y	−24	1E-24	A trillionth of a trillionth
Hella	h	−27	1E-27	A trillionth of a quadrillionth

Plus power means 1 followed by that many zeroes. Minus power means a fraction with 1 on top and 1 followed with that many zeroes on the bottom. The convention is that plus powers are capitalized and negative powers not

zeros, but missing one, right? I mean "bi"llion ought to be 1 followed by two groups of 000's, "tri"llion, 1 with three 000's, "qunt"illion, 1 with five 000's, etc. right? They all seem to be off by one group of 000's. Go figure.

Of course, a good scientific calculator can do all this for you, but I want you to *understand* these things without a calculator, really understand what the answer you get means, and be able to estimate things with just a pencil and the "back of an envelope," or even your mental blackboard which I hope you can develop and exercise while reading this chapter.

People haven't *named* anything smaller or bigger than what's in Table 2, yet. It already covers 60 powers of ten! Of course, there are bigger and smaller numbers. Why would anybody want to measure something as small as 1E-15? Femtosecond lasers put out a pulse that is that brief. During 1 fs, light travels $1E-15 \times 3E8 = 3E-7$ m, or 0.3 μm (0.3E-6), a wavelength of ultraviolet light. Attosecond lasers are being worked on. Can you imagine that? A hydrogen atom electron takes 150 attoseconds to go around its nucleus, so you can see that, with a few-attosecond pulse, you can take a flash photograph of a chemical reaction happening, freezing the electrons with their pants down, so to speak (electrons do the reacting).

Why worry about an Exasecond? The age of the Universe is 0.44 Es. The mass of the Earth is about 6 Xennagrams, or Xg. What could possibly be interesting about a hellagram (hg)? An electron weighs 0.9 hg. The diameter of our Milky Way Galaxy is about 0.9 Zettameters (Zm). The nearest star is about 4 ly (light years) away, and our Galaxy is about 1E5 ly across and 2,000 ly thick. If every star were spaced just like the nearest one is to us, there would be about $\pi/4*(1E5)^2 \times 1E3/4^3 = 2.4E11$ (240 billion) stars in our Galaxy.

Light year? That's just the *distance* (not a time) light can go in a year, about 9.46 Petameters (Pm) (Fig. 2). Sounds funny, right? You can do this stuff! It isn't hard to do even astronomical calculations!

Fig. 2 Petameter (F. Wicke)

Mental Blackboard

Does this sound hard? It isn't, and it's very useful.

Here's a meditation! Go inside somewhere it's not too bright and close your eyes. Imagine your hand out there with a big bright, fluorescent marker and write "12." Write another "12" under that and add them. Don't open your eyes! Can you see 'em? Can you get "24" in your mind's eye below the line? Now erase this and write "12" with a "3" under it and multiply. Can you get "36"? Now, something more complicated: Add "1234" and "1111." Can you do that? Meditate until it comes into focus. Ah, there! Keep at it until it's easy in odd moments this week. People will just think you're meditating or spaced out. In 2 weeks, I want you to be able to multiply 123 by 3.14 and compute $(123)^2$! You can do it. When you're expert at that, you can figure the area of a circle in your head!

My message to you: Don't be afraid of numbers, even if they're very big or very small!! They're your friends and they'll help you a lot. It's far too popular these days to laugh at geeks and think people who are good at mathematics are strange. Here's a whole new country for you to explore, and it's much easier than you've been led to think. All you need to be able to do is multiply a few small numbers, and add.

Read this chapter again whenever you're having trouble with numbers. Rather than dragging you through a whole bunch of boring exercises, I'd just like you to understand everything in it!

Magic Numbers

Why on Earth do we have 24 hours in a day, 12 months in a year, and 7 days in a week? Why not 10 of each? Why do all jokes have three paragraphs? You may *not* have wondered about that, but here we go (Fig. 1).

We have ten fingers, right? I've seen science fiction stories that claim all the 12's (12 hours, 12 months, dozens, 12 grades, duodenum (believe it or not, it was originally called that because its size is 12 fingerwidths), 12 Apostles, 12 days of Christmas, 12 signs of the zodiac, 12 knights at the Round Table, 12 steps, 12 inches per foot …) are because we're descended from Star Beings with 12 fingers. Counting my own fingers, that seems unlikely.

You might guess that the twelves are because the word "months" comes from "moons," and the Moon goes through a cycle about every month. So do women, so it's an important cycle.

But wait! It wasn't always so! Did you ever wonder why the names of the last four months are *Sept*ember, *Octo*ber, *Novem*ber, and *Decem*ber? Did you know that there were once just 10 months? King Romulus invented the *original* Roman calendar in 750 BC. In the *Romulan* calendar, those last four months were named 7, 8, 9 and 10! Just before that were Quintilis and Sextilis. However, because of the Moon cycle, all these months were 30 or 31 days long, so the whole year was 304 days, and winter came when it would. In 700 BC, people woke up and added January and February. Finally, in the time of the Caesars, Quintilis was renamed July in honor of Julius and Sextilis became August for Augustus. This was the Julian calendar.

Because, viewed from Earth, the Moon's cycle takes 29-1/2 days, a year is 10.6 days longer than 12 Moon cycles, so it's a pretty bad approximation to say there are 12 months in a year. The Julian calendar tried to take care of that by decreeing that seven months have 31 days, four 30 days ["thirty days have September, April, June and November"] and then February has 28 except on Leap Year. If you add that up, and average over four years, you get 365.25, which is a genius *algorithm* (way of solving a problem, see *Islamic Science*), accurate within 0.002 % of the right answer of 365.256365. The arrangement of months that we use today is another reminder

Fig. 1 Magic numbers (C. Phipps)

that people were still smart back in Caesar's time. Orthodox Greeks still use the Julian calendar.

So the twelves are because of the number of full moons during the four seasons. And those twelves really get in the way of a more reasonable system which the rest of the world uses, called Metric (see chapter on *The Metric System*).

Now what about the sevens? There are seven musical notes per octave, seven days in the week, seven days for creation, Seven Seals, and—are you ready—seven is the spectral classification of White Dwarfs in the Yerkes Spectral Classification System. Of course, those are stars, not small-size people.

Did you know that seven "is the first natural number for which the next statement does not hold: Two nilpotent endomorphisms from Cn with the same minimal polynomial and the same rank are similar." I'll bet you didn't, and I don't care, either.

So why the seven? Well, that goes back to the Babylonians! During the Jewish captivity in the sixth century BC, Babylonian practices went into the Hebrew calendar and then to us.

Now why did the Babylonians do it? They really liked that number, and thought it had magical powers. So there you have it. Seven Samurai, dance of the seven veils, seven dwarves, seven deadly sins, Seven Wonders of the World, Seven Hills of Rome, all magic numbers.

What about the number 3? The young prince looks into the well and makes three wishes. Third time's the charm! On three, we jump. The Holy Trinity. A trilogy. Two out of three ain't bad! Goldilocks and the Three Bears. Three Little Pigs. A triptych.

> One little maid is a bride, Yum–Yum,
> Two little maids in attendance come,
> Three little maids is the total sum,
> Three little maids from school.
> *W. S. Gilbert & A. Sullivan, The Mikado, 1885*

What's going on here? My wife says three things in a dream are archetypal. Well, the "rule of three" has been around in our tales and jokes since way before "Veni, Vidi, Vici." It's just the way humans like to tell stories, probably for tens of thousands of years.

What about 666? Back to the Babylonians again, who had 36 gods, 3 for each sign of the zodiac (and there are 12 of those, of course!). Add the numbers from 1 to 36 and you get 666. That's gotta be a magic number if you have 36 gods! In the Christian Book of Revelations, the number takes on a more sinister tone, but that's all it was at the beginning: another magic number.

So I've got to tell you a funny story (Fig. 2). This really happened, less than a year ago. My wife and I were about to board the train north from Stockholm to

Fig. 2 The train to hell (F. Wicke)

Säter, where my grandmother left 125 years ago to come to America. The end of the line is Falun. I couldn't help noticing the number of this train: 666. I bumped Shanti with my elbow and pointed at the ticket. She didn't quite get it because they didn't dwell on that stuff in her church. Just at that moment, the young conductor came by. I wanted to be sure we were about to enter the right coach. "Is this correct?" I asked, showing him my ticket and pointing at the door. He didn't miss a beat. "Yes! This is the Devil's Train! This is the train to Hell, and to Falun." You can't believe how hard I laughed. Later, on the train, he came by, leaned over and said, conspiratorily, "Well—it seems to be going all right *so far!*" All the way from the Babylonians to Sweden, 2014.

My point here is that people are storytellers. If they don't understand why something happens, they'll make up a story about it. Magic numbers and incantations play a big role in those stories. Some of those stories are quite beautiful. With science, in the last few centuries of our species, we can explain why some things happen, and invent a number system that's easy to use and makes sense, free from the baggage of the past—that's what the next chapter is all about—but, I hope, without losing all the magic.

Introduction: Weird Reality

This chapter is the introduction to everything else. We'll introduce Lasers, Relativity and other things, highlighting their astonishing aspects, but get into it more deeply later. The main purpose here is to talk about the truly weird things in science. I call the chapter "Weird Reality," because the things in it are not just theories but how the World Really Works, so far as we are able to understand it today. Even if the scientists don't know, in many cases, you will often be able to determine whether something could happen, or not, by using common sense.

When I went to MIT, no-one gave a course titled "What We Don't Understand." I sure wish they had. Such a course would have made many of our lives so much more interesting and productive, because it would have shown us where to look. In this chapter, I want to show what is weird that we do understand, what we only partly understand, what is at the very edge of what we understand and what is still a mystery. "We" "understand" parts of these mysteries, but not all of any.

The Big Bang: Dark Matter, Dark Energy, and … Stuff

In the beginning … the earth was without form and void, and darkness was upon the face of the deep … Then the Spirit said 'let there be light.'

How's that for a lyrical description of creation by a possibly illiterate person (Moses) in Pharaoh's court 3500 years ago? (Fig. 1).

About 13,798,000,000 years ago, there was a huge singularity. Singularity? That's when something goes to infinity in time or in space. In one tiny part of one bubble in the vacuum, our entire universe sprang into existence and began expanding.

In ten billionths of a trillionth of a second, it was about the size of a golfball (or the size of an atom, depending on what expert you are reading!). This is a trillion trillion times the speed of light!

After one second, the size of the universe was about 0.1 light-years, its temperature was 10 billion degrees and its density was 400,000 times that of water.

Fig. 1 In the beginning ... From the BOOK OF GENESIS ILLUSTRATED by Robert Crumb. Copyright © 2009 by Robert Crumb. (Used by permission of W. W. Norton & Company, Inc.) and from GENESIS: Translation and Commentary, translated by Robert Alter. Copyright ©1996 by Robert Alter. (Used by permission of W.W. Norton & Company, Inc.)

That means that in that first second, "it" (whatever "it" was) was traveling at 30 million times the speed of light. Isn't that incredible? This process is called "inflation." But things can't go faster than the speed of light, right? *They can't, until they do!* (Fig. 2)

Why did the Big Bang happen just *there,* at that location? What is a location, if the whole Universe we know and love didn't exist before the Big Bang? What is "here?" And what is "before" it existed? If the Universe has a boundary, what is beyond *that*?

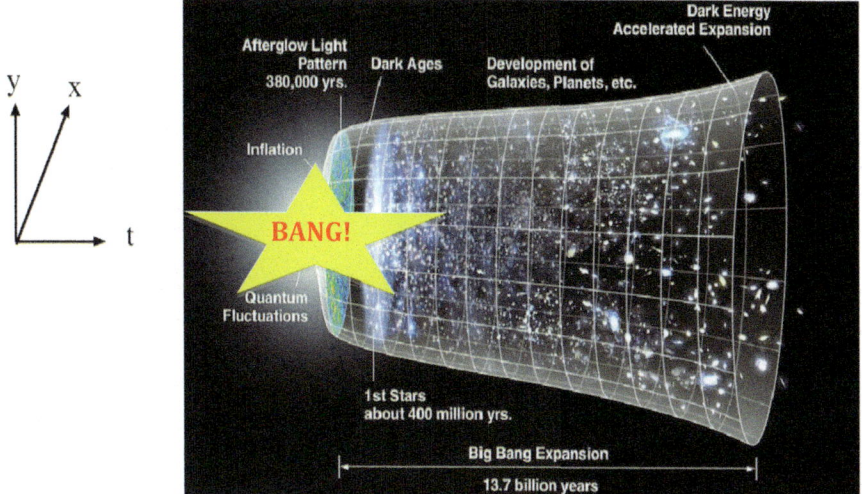

Fig. 2 The Big Bang. In this figure, "*Dark Ages*" refers to the period before the first stars, not a period in medieval history (Modified from NASA public domain)

What if it happened *again* in your bedroom, tonight?

Scientists and theoreticians don't know what the laws of physics allow and forbid in this case. Did you see the movie *Interstellar?* You should.

Did everything originate from a *black hole*? We can only guess. Check out the end of this chapter for more on black holes.

After a few minutes, it slowed down to a normal speed less than the speed of light. Later yet, in our time (the past few billion years), it sped up a bit due to something called dark energy, but never again faster than light. That is what the thimble shape of the object in the figure is trying to illustrate, and why the skirt of the thimble is curved out toward the right, as it goes faster. I know it's confusing: only scientists are used to looking at an *x-y-t* diagram. But we simply can't draw in four dimensions.

That's just our Universe—there are probably a lot of others. No, no, no—after 4000 years trying to prove we humans are at the center of everything in some way—we are *not*. Not the center of the solar system, or the Galaxy, or the Universe, or Reality. Recently, we learned that there is a planet pretty much like Earth called Kepler 186f, 500 light-years away. If you're paranoid, that's not a big concern, because they couldn't get here. That's a little over 400 million billion kilometers, and it would just take too long. *Unless they used wormholes!* But, it's ridiculous to think there is no life except on Earth. We are not unique!

Did you know that *only 5 % of the mass of the Universe is normal Stuff like you, me, the cat and the stars* (baryonic matter). The other 95 % is dark matter, radiation, "dark" energy and whatever else. Right now, we believe 27 % is dark matter you can't see and twice as much (68 %) is dark energy. There are 20 times more of the Stuff you can't see than of the Stuff you can! (Fig. 3)

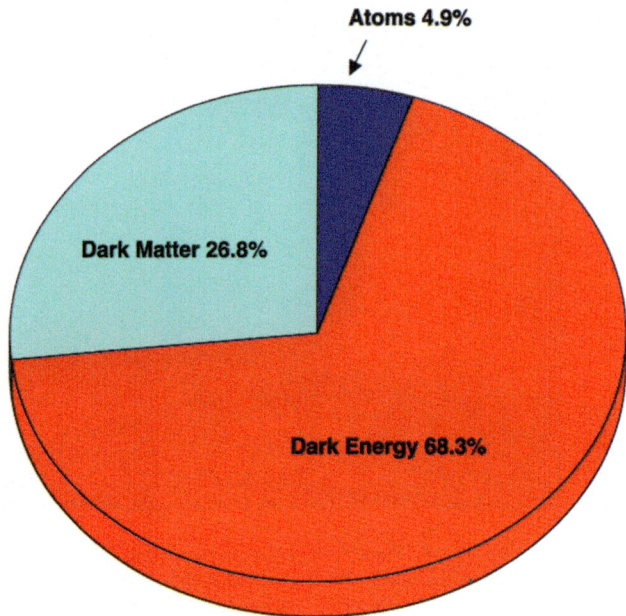

Fig. 3 Budget of the universe (C. Phipps)

Dark matter is not necessarily mysterious. It could just be rocks, dust, dark galaxies, millions of black holes or brown dwarfs—stars that are too small to really light up. It also could be weird particles that we can't see, like axions, neutralinos or great big, heavy imaginary particles called WIMPs (weakly interacting massive particles) that haven't been discovered yet.

Dark energy is another thing altogether. From the figure, you can see it's a huge part of our Universe, but *nobody knows what it is!* One scientist has called it "a placeholder for our ignorance," but we know *something is making the universe expand faster the farther it is away from us*. That means the more the universe expands, the more dark energy there is. It could just be vacuum energy, like we talk about in the section called "ZPE" at the end of this chapter.

What Is Stuff?

This is a rather philosophical question to throw into the middle of this book—but … what *is* stuff? By this, I mean physical reality that you can see, feel and taste.

You may have heard that things (including us) are made up of molecules, which are made of atoms, which are made of electrons racing around nuclei. A hydrogen atom nucleus is about 30,000 times smaller than the atom. Its one electron zooms around it at about 2000 km/s, fast enough to get halfway across the USA in a second. One trip around the nucleus takes just 150 as (attoseconds: billionths of a billionth of a second)!

Fig. 4 Atom of oxygen
(F. Wicke)

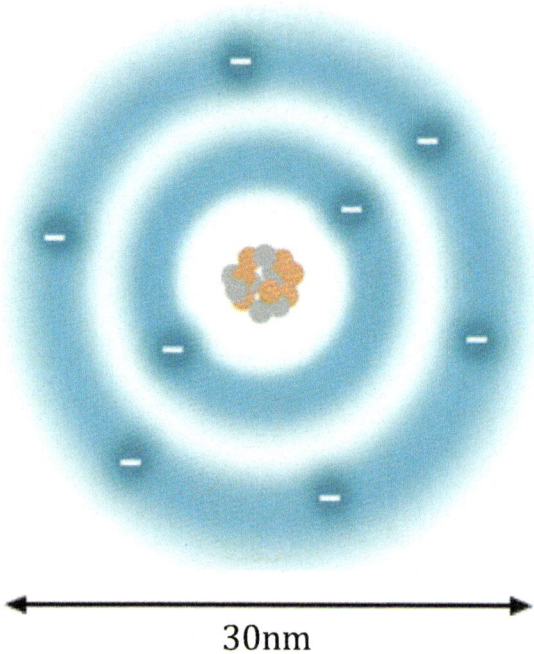

30nm

In Figs. 4 and 5, I've shown an oxygen atom, plus two of these hydrogen atoms linked up with an oxygen atom to make hydrogen suboxide, H_2O, which we know as water.

Human cells are made up of many different kinds of molecules (Fig. 6). None of these illustrations are to scale. In reality, an atom is mainly empty, except for parts that take up a few millionths of a billionth of its internal space. It's useless to express that in percents. Early scientists imagined these tiny bits to be analogous to the sun and planets in our solar system, with the negative electrons held away from the positively charged nucleus by centrifugal force against its electrical attraction, in the same way the planets are kept from crashing into the sun due to gravity.

See *Lasers* for a better understanding of the atom.

For scale, imagine a hydrogen atom nucleus blown up to the size of an orange. Then the atom would be 2 km (1.4 miles) in diameter, with all the space in the middle empty.

But that's a false picture! The electrons, protons and neutrons in an atom are not solid bits of *anything*, just dense arrangements of electrical and other forces, and what we call their *size* is just a scientific definition. Protons and neutrons are made up of other things called quarks and these are held together by things that are like photons, but not quite, that you never can see. Reasonably enough, these are called gluons and they *pop in and out of existence* as they transmit the forces that hold things together at that level.

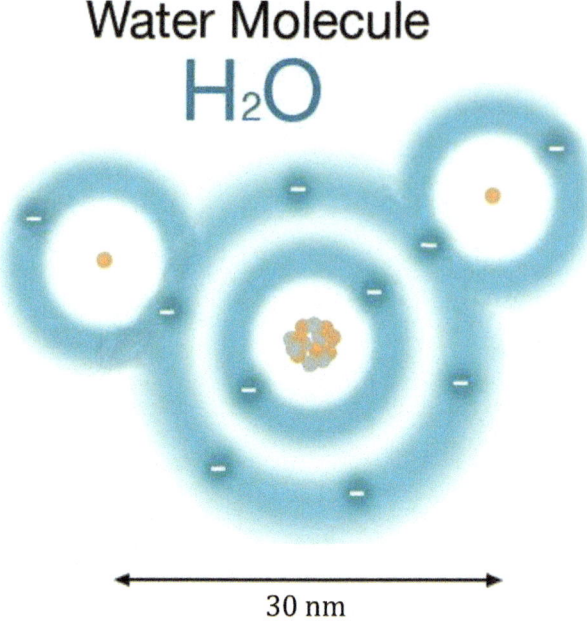

Fig. 5 Water molecule (F. Wicke)

Fig. 6 A human T-cell is made of molecules, which are made of atoms, whose orbiting electrons push on things (U.S. National Institutes of Health, public domain)

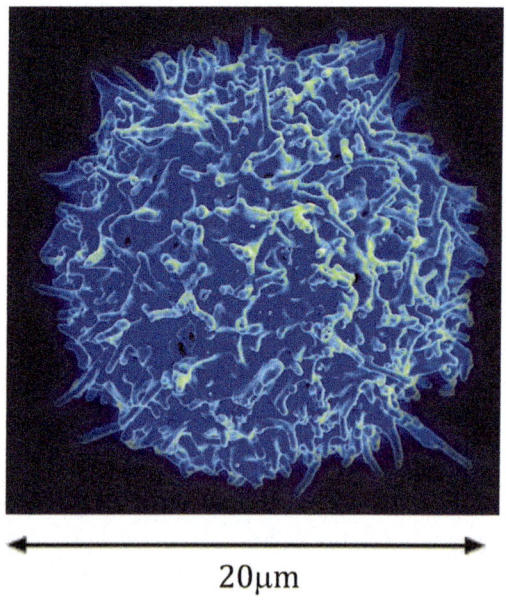

What Is Stuff?

So far, we don't think electrons are made up of anything more basic. But that's been said of the parts of an atom many times before!

To sum up, an atom (and therefore you) is like Gertrude Stein's famous pronouncement about Oakland, "there's no there there." And since all the stuff (including you) is made of atoms, there's no there there, either!

When you do a pushup, or hold your forehead in your hands late at night, you are feeling electrical forces. You're pushing on the electrons in all those atoms that make up the molecules in the cells you are made of.

By the way, in my case, that's about ten trillion quadrillion atoms (80 kg). Just so you appreciate the number of things that are cooperating to keep you alive! But, if it's all electrical forces, where's the *stuff*?

F. Wicke

Answer: there's *no stuff*. Just forces. *The force fields have energy, and energy has mass.* Remember $E=mc^2$? This is the only equation you will see in this book, and I want you to get a feeling for what it means. Einstein said mass is equivalent to energy, and energy to mass. Energy divided by the speed of light squared gives the mass of everything ($m=E/c^2$). *There isn't anything else!* Now, c^2 is a pretty big number, so if you could instantly convert your coffee mug (1 kg) into pure energy, you'd have a 20 megaton bomb and pretty much wipe out the city you live in, just to put $E=mc^2$ into perspective for you. In beginning algebra, I told my teacher about that, and she was amazed. The hydrogen bomb only converts about 3 % of its mass into energy (Fig. 7).

Imagine that you're bringing a bunch of tiny little negative charges together to make a single electron, like the guy in Fig. 8, and that they all add up to one electron charge. This is a thought experiment: you can't divide up an electron charge into hundreds of tiny pieces because of quantum mechanics.

When they're far away from each other, you won't feel a thing. As you bring them closer, they will push back on you because like charges repel each other. So now it takes a little work. The closer they are, the harder they push back and it takes even more work to bring them closer. You've done this sort of thing when you push two repelling magnet poles together. When you're all done, at a diameter of 2.8 fm (femtometers—millionths of a billionth of a meter—see *Metric System*), you will have expended 511 keV (kilo electron volts) of energy and that is exactly equal to the mass of an electron when you get all the units right. The energy of the electric field of an electron *is* its mass! There isn't anything in the center either, *except a weird singularity we don't understand!* Same thing with protons and the other parts of matter.

The electron's electric field goes to infinity at its center. To deal with that, scientists draw an imaginary circle around the singularity at a radius that just gives its measured 511 keV mass when you cram the charge inside that radius, and call it the electron radius. Same for a proton.

Fig. 7 Twenty megatons from your coffee cup (DoE public domain)

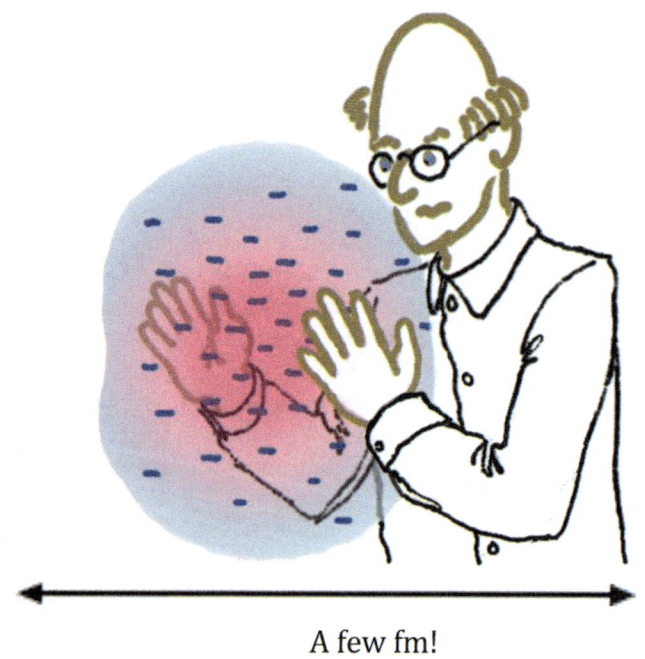

Fig. 8 Forcing charges together (F. Wicke)

Fig. 9 Pushup (US Government public domain)

Anyway, when you're doing a pushup like the young lady in Fig. 9, what keeps your hand from pushing right into the floor and becoming a solid piece together with it? Mostly dirt and roughness. You know how tape doesn't stick if it's dirty, right? When things stick it's because the atoms intermingle. Geckos can walk across the ceiling by sticking to the atoms in it. And two very flat pieces of stainless steel or glass will stick together.

What does all this mean to you in a practical sense? That's up to you to decide. It does help to make you wonder what's real. In the 1960s, I used to hear that question a lot. It was a time for questioning reality. My answer used to be, "Walk out in the middle of Highway 101 in a dark suit at midnight, and you'll find out!"

What *is* real? It's a mystery. It's up to you. Your own experience is your only guide. Your experience will be different from mine. *"Je pense, donc je suis,"* (I think, therefore I am) as the French philosopher and mathematician René Descartes said about 400 years ago. What others tell you is a second level of reality, not as real as your own. What you hope for in the future based on lots of belief and philosophy is a third level. When we cease to exist in this form, we may come to know yet another level.

If you saw the movie "Her," you may have wondered seriously about reality. Theo is in love with an operating system (OS) called Samantha. And, in terms of intelligence, feeling, memory and so on, she is more than his equal. In fact, she's also in love with 634 other people. But she doesn't have a body so she can't touch him, or he her. Is she real? Is Samantha's life what it will be like for us after death? I don't know. No one who has been there has come back to tell me.

But, you say, I *see* stuff! I feel it, breathe it, eat it! Yes, you see it, because of the colors and sparkles of light bouncing off it or shining through it. Or, perhaps, coming from it if it's glowing in the dark. That's because electrons in atoms absorb light or other energy and send back some colors better than others. You feel it because cells on your fingertips send electrical signals to your brain when electrons in the orbits of their atoms in the molecules of their cells push on something. Or when wind molecules brush your face. Gravity pulls on mass, so my coffee mug has weight and stays put on the table.

Gravity

F. Wicke

Speaking of *gravity* ... *what is it*? We take it for granted and deal with it all the time we're alive, but we don't know what it *is*. When we're on the Earth, it pulls us down toward the center of the Earth. If you were on the Moon, it would pull you toward the Moon's center.

If you think about it, that must mean there's no pull at the center of the Earth, because all the Earth's mass is around you instead of under you, right? It's true—you'd be floating around like an astronaut. But there would be other problems in the middle of solid iron at 6000 °C.

What is gravity and why does it pull things? Let me say this right now: nobody knows! Oh come on, you say, of course they do! Well, yes, people have developed abstractions by which they can make sense of it. Cosmologists talk about curved space. We are familiar with electrical and magnetic forces, so we can imagine gravity is something like that.

But gravity *pulls on everything and repels nothing*, while electrical and magnetic forces can either attract or repel. Electrical and magnetic forces travel at the speed of light, but does gravity? Einstein thought so, and he was probably right. If an object suddenly disappeared, would everything else in the universe instantly feel its absence, or would it take a few years to feel that loss if you were a few light-years away? Probably, there'd be a delay, but we don't know how much. In spite of looking with very delicate measuring devices for half a century, no one has yet measured the speed of gravity.

Because gravity is a weak force, it's hard to measure, and it's fiendishly difficult to move something around fast enough to see if its pull is felt somewhere else instantly or just at the speed of light, or some other speed. The "feel its absence" experiment makes sense: see if you can measure anything after a supernova explosion (Fig. 10). But, wait! After the explosion, you would still have a spherical, expanding thing with about the same mass, centered at the same place, even if more of it were energy (remember, energy is mass, too!) So you might not measure anything.

Another idea: see if there's a delay between the radio pulses of a twin pulsar (Fig. 11) and a gravity pulsation you might pick up from it. But all these ideas have so far been in vain, as far as measuring the speed of gravity goes.

Fig. 10 A supernova (NASA public domain)

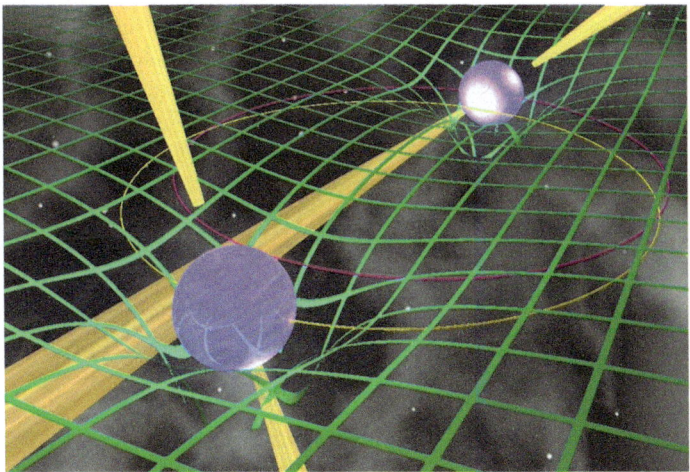

Fig. 11 Twin pulsars. (Prof. Michael Kramer, Max-Planck Institute for Radio Astronomy, Bonn by written permission)

F. Wicke

We do know gravity waves must exist, although that's a long, long way from measuring the speed of gravity. How was that done? The 1993 Nobel prize in physics was awarded for doing the experiment I suggested in the last paragraph. Forty years ago, two scientists called Taylor and Hulse used a radio telescope to show that the time it takes a certain pair of pulsars to rotate about each other was decreasing by—get this: 75 millionths of a second per year! That's a really fanatical pair of guys you say? But that 75 μs was just enough to match Einstein's 1916 theory of how much energy such a thing would radiate in gravity waves, and that's the only explanation for the slowing they saw!

This just in! (March 17, 2014): Gravity waves, or at least their effects, *have been seen*! John Kovac and his team at the Harvard-Smithsonian Center for Astrophysics have seen spiral ripples in the polarization of the cosmic background. It's too hairy to explain here, but their results, measured at a telescope in the Antarctic, show that gravity obeys quantum mechanics. Gravitons exist! This statement has not had any proof until now.

Woops! ***This** just in!* (January 31, 2015): what they saw was just dust. Oh well.

Now, all we have to do is wait, because two giant experiments that are just turning on in Louisiana and in Washington State will see these waves any day now. See the very end of *Lasers*.

Umm ... "Cosmic background"?(!) Back in 1964, Arnold Penzias and Robert Woodrow Wilson were looking at the sky at microwave wavelengths. This means radio waves with "lengths" between their crests of centimeters, which in the days of radio were "micro," and in this case a frequency of about 160 GHz (billion cycles per second). They saw noise that would not go away, and looked the same whatever direction they pointed their antenna. Robert Dicke suggested this might be the cosmic background radiation, radiation left over from the Big Bang, the very oldest light in the Universe, light that has been so "redshifted" because of the Doppler effect (frequency dropping as things move away from you) and relativity that it looks like radio waves. This was a *brilliant* idea! (See the chapter on the *Blue/Black Sky* and the "relativity" section below to understand this better). Penzias and Wilson—but not Dicke—got the Nobel Prize for that. Go figure.

What DOES the "acceleration of gravity" mean? Let me tell you a true story, and then you can go do the experiment yourself. Years ago, I drove to Palo Alto in my Volkswagen to fill a bunch of helium balloons for my birthday party (Fig. 12). We lived in the country half an hour away from Palo Alto, so either they would have to be filled already or I'd have to buy a helium cylinder and I didn't want to do that. But I had balloons and I knew where I could find helium at Stanford. I filled about a dozen balloons, attached strings, pushed them one by one into the back of my VW and took off. I almost had a wreck! The minute I accelerated, they all crowded forward in front of my face! I couldn't see! Panicked, I stepped on the brakes—and all the balloons obediently returned to the back! I pulled over and thought about that for a minute. What was going on? Then I realized: helium balloons don't care a bit

Gravity

Fig. 12 Helium balloons in a VW Beetle (F. Wicke)

where the pull comes from—gravity or acceleration. Both are the same to balloons and they move away from it! Gravity pulls you down but the balloon goes up. Acceleration pushes you back but the balloon goes forward. When the car turns right, the balloon goes right. When the car turns left, the balloon goes left, as if it knew in advance which way to go! Same thing! You can do this experiment yourself, but please have just *one* balloon and have someone else drive! Is there any difference between gravity and acceleration? No.

Oh—and Isaac Newton did *not* have an apple fall on his head. But he did know that, except for air resistance, big and little things fall at the same rate. Galileo dropped some big and little things off the leaning tower in Pisa to prove that. In a vacuum, a feather drops just like a rock. It is not at all obvious that this should be so. Newton went on to invent a whole new mathematics (calculus) 400 years ago in order to calculate how the planets move in their orbits in the sun's gravity, and moons about planets in theirs. This is an example of things falling at the same rate. What he found was: the pull of gravity depends on your mass (you knew that, I hope: if you're twice as fat, you're twice as heavy). It also depends on your distance from the center of the object that's pulling you (Fig. 13). Newton's genius discovery was that the pull of gravity is a certain number times your mass times the mass of the planet or object that's pulling you, divided by the distance to the center of it *squared*. If I climb a very high mountain, gravity is a little weaker because I'm farther from the center of the Earth. This number is the same number everywhere. Newton made a guess about how gravity worked and he was right! The relationship between the length of the "year" of a planet and its distance from the sun proves it.

Fig. 13 All things attract all other things with their own gravity (F. Wicke)

Why did I say "planet *or object*" there? A big mountain pulls sideways on you with a measurable force. You pull on your friend, too, each with your own gravity. Did you know that? You'll never feel it. If you're both about 80 kg, and stand 1 m apart, the force is about a millionth of an ounce. The mountain's pull? Maybe a few hundredths of an ounce. Gravity is a weak force. At the top of Everest, I would weigh 0.2 kg less, and that would be measurable with a really good bathroom scale.

Inertia

Is inertia the same thing as gravity? Someone asked me that just last week. Is there a difference between pushing against gravity and pushing to get something moving against inertia?

Why is it, really, that you have to push hard on a big thing to get it moving, and again to get it to stop? Imagine you have a 1000 kg railroad flatcar on frictionless steel rails and want to get it moving just 5 miles/h (223 cm/s). Let's say it has perfect bearings, so there's no friction at all. You could do this by pushing on it with a force of 222 N (about 50 pounds) for 10 s. During the push, you put 2.5 kJ of energy into the car. There it goes down the tracks! It looks exactly the same, and doesn't glow or anything, but it now has 2,475 J of energy in it that it didn't have before, and you're a little out of breath (Fig. 14).

Fig. 14 Railroad flatcar and a bag of potatoes (F. Wicke)

Now say I'm sitting on the car while this happens, with a 20 kg sack of potatoes, and you push a little longer to get the car and me and the potatoes going. I won't feel any different afterward, but I (80 kg) will have 200 J that I didn't have before, and the sack 50 J, from your point of view. In my frame of reference, none of that is true. The sack and I are not moving.

Relativity at Low Speeds: Kinetic Energy

Now, this is getting twisted, but say I throw the sack back at you at 5 miles per hour. *I added another 50 J to it* to do that but to you the sack now has no energy at all, because it's standing still (after it drops on the tracks). We added 50 J to it twice, but the only change for the sack is that it's further down the track. *What is going on here?* What *is* this *kinetic energy*, and where did it go? How come I can push all day against the seat of my chair and get nowhere? Is there any difference between *those two forces*? No. A force is a force. And, there's no such thing as negative energy except for tiny, tiny times. But if I'm constantly pushing against gravity, am I adding energy to the planet? No, *unless I climb some stairs*. Force acting through distance is energy. It doesn't take any work to just stand there, except for the way our muscles twitch.

Still, what *is* kinetic energy? Answer: at normal speeds, it's kind of a bookkeeping exercise that keeps track of physics in *different frames of reference* moving relative to each other. Isaac Newton figured that out about four centuries ago.

Einstein was here

F. Wicke

That's not satisfying, is it? Look very hard, but you won't find a different answer in most beginning physics books. But remember $E=mc^2$. The truth is, you put energy into that flatcar, into me, and into the potatoes. Therefore, *you changed our **mass**, very very slightly, according to you, while, according to me, there was no change. That's where the energy went!!* This is relativity.

Relativity at High Speeds

At very high speed, closer to the speed of light, you (stationary) and I (moving) will disagree majorly about mass, time, velocity, distance and energy, as Einstein figured out about 100 years ago in his *special theory of relativity*. As I near the speed of light, you will think I have a tremendous mass, but I will feel normal.

And, you will age much faster than I do!

There was a *Twilight Zone* story based on that in 1964, entitled *The Long Morrow*. Our astronaut must travel to a planet that is 140 light-years away. Because of relativity, the round trip will take only 40 years back on Earth, because he'll be going very fast. But, he's in love with a young lady on Earth and she will be 40 years older when he comes back (Fig. 15). The young lady goes into a suspended animation unit in order not to have aged when he returns. But, on the spaceship, Commander Stansfield turns off his suspended animation unit in order to look the same age as she when he returns. It's a sad reunion. Even sadder because Rod Serling got his relativity wrong! Because of his high speed, even *without* suspended animation, he would return only 19.2 years older. In contrast, she would have aged 277 years and I don't think she'd make it, even with suspended animation. But it would have helped alleviate boredom for him. Which brings up:

Time Travel

In many of the science fiction stories we all love, you're supposed to be able to travel backward in time. But that won't work, and you can understand why just from logic.

Backward Time Travel

If you could go backward for more than some attoseconds, it would lead to a terrible instability, a *contradiction in reality* that would tear the fabric of time, or create an alternate reality from which you, the time traveler, could not escape. I'm not sure which one of those would happen, or which you would like least. This problem is outlined in the wonderful movie "Back to the Future."

F. Wicke

Let me illustrate. Some really wise science fiction writer (whose name I can't recall now) drew my attention to the problem in a short story I read late at night half a century ago when I should've been preparing for exams. Someone goes back in time and moves one grain of sand on the beach near present day Manhattan. In the future, the Empire State Building disappears. But of course much more than that has changed. Maybe all the people who worked on that building did not have jobs, or perhaps they ceased to exist. Why would that be? It's the butterfly effect amplified like the screech of the sound system in the chapter on *Exponentials*. When that grain of sand was moved, it led to a change in the ocean flow which left part of the island of Manhattan underwater a million years later, so the place where that building stands was not there. Of course, it's also possible for that instability *not* to happen so that no one would notice in the future, but that's not the point. Something would have happened that did not happen.

Why is that? What I'm describing is yet another example of a feedback loop. There are two alternatives:

Option 1: Imagine that I go back a century in time and am very careful to just observe, not to disturb a *thing*. I cross a dusty Main Street one afternoon without looking, causing two model-T's to collide. One of the drivers, who happens to be my great grandfather on my father's side, is killed. Now I do not exist. The universe we know would now be locked in a logical contradiction and some part of it would implode. A whole sequence of alternate universes, or contradictions, branches out from that one event like a tree. My grandfather and father and his sisters and their many children, and their children, also don't exist in the present, nor do the effects of all the billions of actions and changes each of them made while actually living. Houses and other buildings vanish. Animals die. Bank accounts vanish. The disappearance of one bank account causes the bank's computer system to lock up. A unit at an oil refinery explodes, because my father wasn't there to anticipate a fatal problem. Most important (to me), I am either dead *or* alive in the present, Schrödinger's cat notwithstanding. I can't be both because my life has been entangled with the lives of so many, many others in all the things I've done, however inconsequential. And yet, in the past, I have ceased to exist. This is not possible.

Option 2: Or, if reverse time travel is possible, a parallel universe must be created in which this death could happen, but the victim was not my great grandfather. Yet, I could never escape from that alternate reality and go back to those who loved me. People in the present would just assume that my time-travel experiment had gone awry. That is the only way things could be left in good order. God keeps good books!

Option 3: Stephen Hawking suggests I could go back if in the past I lack free will ...

I expect that backward time travel is so improbably energy intensive (for the Universe!) that that is why it doesn't happen for more than attoseconds.

However, it does happen for these tiny times in the quantum world (check out Feynman diagrams)! Not only that, for even large clouds of atoms these days, it's possible to be in a state of existence and non-existence at the same place, if it's cold enough that thermal vibrations don't mess up the quantum state. Recently, Alexander Gaunt and his team at Cambridge made a so-called "Bose–Einstein Condensate" that contained 100,000 atoms. The experimenters had to make a refrigerator operating at 40 billionths of a degree above absolute zero to do that.

Forward Time Travel

As long as we're on the subject, what about forward time travel? Well, that's easy!

Which is why I brought it up with that *Twilight Zone* episode. It's easy and you could do it tomorrow, if you had the right spacecraft. Spacecraft already move slowly into the future because they're moving fast compared to you, in a gravity field, and your GPS has to take account of this effect. Of course, it's only microseconds. Satellites zooming around the Earth with accurate clocks on them have confirmed Einstein's theory about that, exactly, even if it is a matter of microseconds a year.

So, to forward time travel, all you have to do is take a very high speed round trip. You get on a spacecraft and accelerate away from the Earth at one G (one earth gravity) for a little over 3 years (Fig. 15). That's comfortable. Everything feels normal to you. Then, you decelerate for 3.6 years, circle a star 19.7 light-years from here and return. To come back, you accelerate at one G for 3.6 years, then decelerate at one G for 3.6 years and land back on Earth. You think you've been gone a total of 14.5 years.

Indeed, the clocks on your ship, and the aging process both say you are 14.5 years older, but all your friends and relatives are 80 years older, and that likely

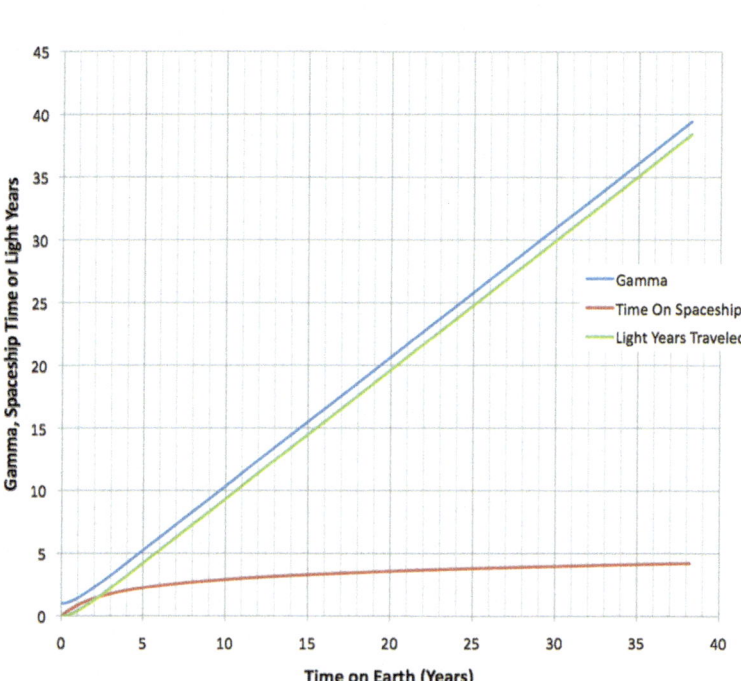

Fig. 15 Relativity. This is for accelerating at one Earth Gravity. Pick a trip time measured on Earth (*horizontal axis*). The *red line* (*use the vertical axis*) shows how many years you think it is since you started on your spaceship, the *green line* the total number of light-years traveled, and the *blue line* gives Gamma. Gamma tells, at a particular moment, how many times your mass has increased, and how many times *time* has slowed down for you (C. Phipps)

means every one of them is dead. It might be frightening and terrible for you or, if you don't like those relatives and wish you could skip forward in history to a time when you imagine all this craziness will be over and see what new things have happened, welcome aboard! This trip is just for you!! Incidentally, you topped out at 99.9 % of the speed of light, and the propulsion system had to provide *1.7 gigatons* of energy to get just you to that speed, let alone the whole rocket and whatever fuel it had left. Then, you have to do it again and again and again in order to get home. That is the only problem with forward time travel. But conceptually it's easy. Certainly not forbidden by any laws.

F. Wicke

But—what if you *didn't* want to go home! You *could* go on like that forever until you ran out of food and water. Time travel really pays off at high speed. After accelerating for 5 years, you'd be going 99.994 % of the speed of light, and time would have slowed for you by a factor of 71. You would have already traveled 87 light-years and you'd be just now zooming past the star Eridanus in Orion. You would have passed Sirius a couple of years ago if you had gone in that direction instead. If you turned off the motor at this point and cruised, you'd be traveling 71 light-years per year!! Also you'd weigh 5.7 t, but you wouldn't feel it in any way.

Check out Fig. 16! It shows how much time is distorted by the speed of the spaceship. It shows that, when you get close to the speed of light, the time

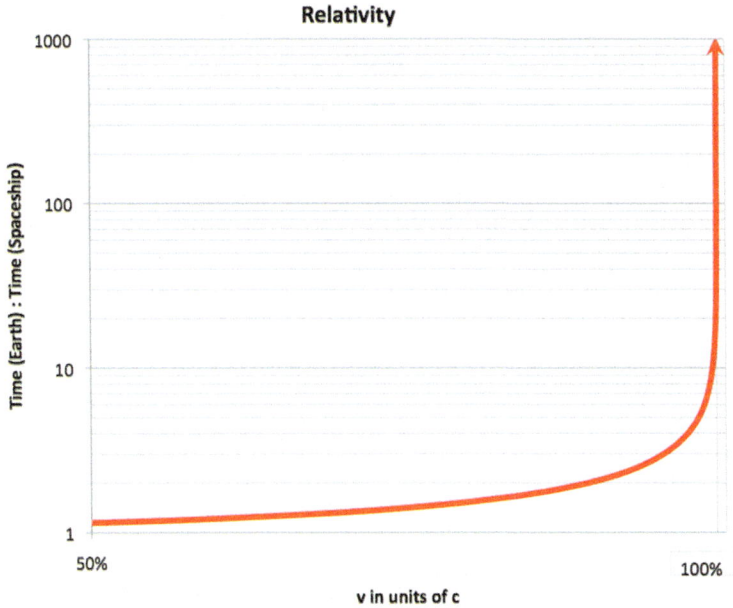

Fig. 16 Ratio of time on Earth to time on the spaceship versus its speed (C. Phipps)

Fig. 17 Space cowboy. From our point of view, his hourglass is stuck! (F. Wicke)

ratio becomes a singularity! At 100 %, it would actually go to infinity if I could plot that.

What does *that* mean? Think about it: if I am sitting on a light wave (which goes at 100 % of the speed of light!), and you are back on Earth, time stands still for me relative to you! Relative to you, I live *forever*. Generations come and go on Earth, but I am the same age forever. God is light? (Fig. 17).

Why doesn't light take infinite energy to accelerate to the speed of light? It's because light doesn't have any mass. (But it does have momentum! Go figure.)

Now, think about this: if the Earth appears to me to be accelerating away from me but to Earthlings I appear to be accelerating away from them, is there any difference? Isn't everything truly relative? No. It is I who have put gigatons into the kinetic energy of my mass, not it.

Einstein dreamed up Special Relativity during his day job at the Swiss patent office. Moral of the story: hang on to your day job!! You might discover something important, especially if the job's not too demanding, and you'll always get paid!

Teleportation

Teleportation (Fig. 18) is just the idea of taking something from one place and reassembling it somewhere else. We all know about that because of *Star Trek*. It does need a lot of energy to do that, because of $E = mc^2$. Remember the 20 Megatons from my coffee cup? And that was just 1 kg. But, it works at the speed of light, and that would allow us to travel at the speed of light without the hazards of going that fast.

Teleportation 41

Fig. 18 Teleporter
(Konrad Summers Creative
Commons license)

Teleporting would have advantages, if it were possible, because the alternatives are so bad. If you're Capt. Kirk zooming along at 99.9 % of the speed of light (*forget* about hyperdrive for now), relativity says that ordinary light looks like X-rays. Each 1 mg speck of dust in the cosmos hits your ship with 480 t (tons!) of TNT. You'll never make it, at least not without a "Bamford Shield," a magnetic field out in front of your ship that ionizes and deflects things.

Teleportation is possible *now*, one state of an atom (so far, not even a whole atom) at a time. It's still ridiculously impractical for people, half a century after *Star Trek*.

Why does it necessarily take energy to go from being stationary in one spot to being stationary 113 m down the track? Obviously, it doesn't if I perfectly recover the kinetic energy at the end of the trip. This is the idea behind Prius, which turns kinetic energy into battery energy when you slow down.

Or, from one side of the Earth to the other? For years, people have been talking about drilling a tunnel through the Earth, sucking the air out of it and dropping a shuttle from, say, New York to Beijing (Fig. 19). *All such itineraries would take just 45 min!* A roundtrip would take 90 min, the same time it takes a low-Earth orbit satellite to circle the Earth. "In principle" (famous last words), this trip would take no energy at all, since gravity itself would provide the energy recuperation. That sounds very impractical too, but you might have read that Elon Musk is planning a "Hyperloop" to take folks between San Francisco and L. A. in 35 min.

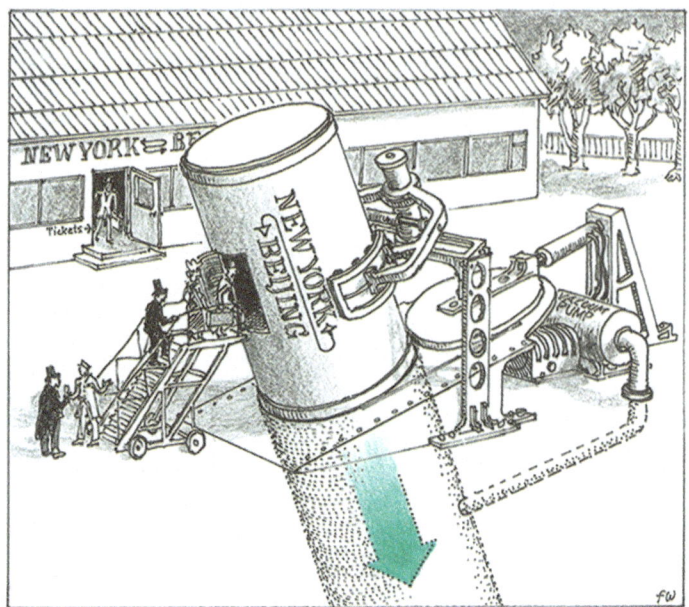

Fig. 19 The New York–Beijing Express (F. Wicke)

This design wouldn't need vacuum, and would use energy, but at least it's a start. Musk's shuttle would just suck air in the front and blow it out the back, traveling on air bearings.

What about teleporting from one side of our *Galaxy* to the other, a trip of 100,000 light-years? That would take an immense amount of energy, of course. But, if you succeeded, it would be 100,000 years from now, long after you died. Did you think about that when you were watching *Star Trek*? That's a very large amount of time travel. Ordinarily (in science fiction movies), the first person disappears in one place and reappears in another. But if it's just an electronic transmission, could it be recorded? If it could be recorded, it could be repeatedly transmitted. Could you reproduce yourself in several places, at different times, and live forever that way? Obviously not. What would happen if those people met each other? That's another contradiction that makes multiple You's impossible. You will be definitely gone here, and you need to hope they can reassemble you correctly at your destination!

Quantum Mechanics: Weirdness on a Small Scale

If you want to worry about what is real, think about photons and electrons.

Light is made up of bits called photons. We know that because you can shine all the light you want on a piece of photographic film and it won't expose if it's too red. That's why darkroom lights are red. Particles of bluer light have more energy and there is a minimum amount that is needed to cause the chemical reaction in film that makes an image. Another more complicated effect like film exposure originally gave Albert Einstein the idea of photons in 1905. He got the Nobel prize for that.

Electromagnetic Waves

Photons are *Electro-magnetic Waves*, just like radio, but at a much smaller wavelength. This kind of wave has peaks and valleys where the electric fields are strongest and weakest, and they always go at the speed of light everywhere in the Universe, so far as we know. Where two waves come together and a plus and a minus field happen to coincide, they cancel. It's not quite the same thing, but you've seen it on a pond when you throw in two rocks: the ripples add or subtract, so you get something or nothing or something in between when they collide. Shine a beam of light of one color on a pair of slits, and behind that you'll see a pattern of light and dark stripes that proves light is a wave, because the pattern is due to the waves adding or canceling each other (interfering). If they were just particles, you'd get a smear, like Fig. 20.

For a long time, physics students were taught not to worry about what a photon *is* and just accept that it can sometimes be a wave or a particle depending on the experiment. Grad school was sort of a contest in your ability to suspend practical beliefs and accept strange new ones, like a modern New Age encounter.

F. Wicke

If light is going through only one slit (Fig. 20), it'll make a kind of blur on the screen (with a bit of structure outside it not shown). But: turn the light down until only one photon at a time is at the pair of slits and *send it through a pair of slits*. After enough photons have gone through to expose the film, you'll get the pattern in Fig. 21, an "interference pattern." But you can't get that pattern unless that photon went through both slits so it could interfere with itself. Yet, it's indivisible! If we want to get deeper into it, the usual story

Fig. 20 A light beam shining on one slit spreads out and makes a blur on the screen (F. Wicke)

Fig. 21 The green gremlin lets the same light beam shine on two slits, and it goes through both, making a stripey "interference pattern" on the screen (F. Wicke)

is sort of a Buddhist koan, where we are asked to hold contradicting ideas. A photon is indivisible but it must interfere with itself to make that pattern, so it must have gone through both slits! *Or*, it was marked, leaving some kind of scent for the others, telling where it had been, like Monarch butterflies migrating.

It turns out that we can get the same effect if we shine a beam of *electrons* or even *atoms* on two slits (Fig. 22). These are *clearly* particles, right? What does it mean if this basic constituent of matter also acts like either a particle or a wave depending on how it feels?

An electron or an atom is *also* a wave with a frequency proportional to its kinetic energy. It's not as weird as some people claim. Actually, each single particle does land on a spot. Only after a lot of particles have come through do they add up to the two-slit pattern. One wonders if it's the same with photons.

Suppose there are detectors sensitive enough that you could actually find out *which* slit a particular photon or electron went through. According to quantum physics, the instant you do that, the interference pattern disappears! *What you decide to do determines the outcome of your experiment*!

This result has been accepted as gospel by generations of grad students, *but there is no publication where this experiment has actually been done and the results reported*. Stephan Ritter and his colleagues in Germany recently figured out a very complicated way to tell that a single photon had passed without destroying it. But, he warns, "you've pulled information out of it, so you do wind up affecting it." As of this writing, whether it could still interfere with photons from the other slit remains to be determined.

Figure 23 shows an actual interference pattern from sodium light.

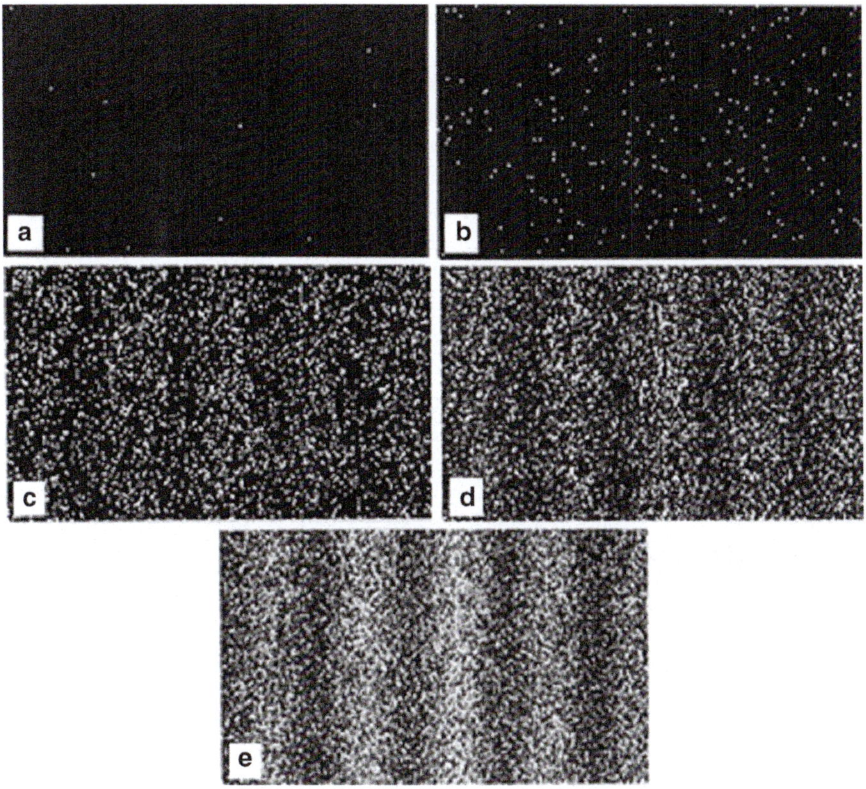

Electromagnetic Waves

Fig. 22 Electron interference adds up (Belsazar Wikimedia Commons)

Fig. 23 Interference pattern. Light and dark stripes from shining a light on two slits, make a pattern of light and dark bands on a screen, depending on whether the two beams cancel or add (Epzcaw Wikimedia Commons)

Quantum Erasers: True Weirdness

A single photon leaves its scent, is marked in some way, by our measuring which path it takes. If it weren't marked, it wouldn't know not to make that interference pattern and make the blur I mentioned earlier.

This just in! (Well, within the last decade anyway): There is a way you can unmark it, *either before or after* it hits the slits, and restore its ability to make that pattern! This is called the Delayed Choice Quantum Eraser and it allows *you* to decide whether to measure which slit it went through *after* it has already gone through the slits *and either interfered with itself or not*! Of course, we're talking about very short times here, but *something that happens now can be made to go back and determine what happened at an earlier time!* You need entangled photons to do that. It's also a feature of "Feynman diagrams," which the physicist Richard Feynman invented to describe how subatomic particles interact. For example, a positron is an electron moving backward in time!

Entangled Photons: More Weirdness

F. Wicke

You've probably heard of these guys. There are lots of stories these days about quantum keys for encryption. Entangled photons are just a pair of photons (there can be more, but let's stay simple) created at the same instant with opposite properties. For example, one of their waves might be rotating left and the other to the right. Let's say I send one of them to a friend using an optical fiber. Neither of us knows which way my photon is rotating, until one of us measures it. If my friend tells me his is right circularly polarized, I instantly know mine will turn out to be left circularly polarized when I measure it, even if our labs are hundreds of km apart.

If you want to be amazed, think about the fact that the state of my photon was *determined* by my friend's measurement!

This doesn't violate Einstein's rule that neither mass nor information can go faster than the speed of light. But knowledge can! (For more about right and left circular polarization, see *Electromagnetic Waves*).

Uncertainty

Normally, physics courses go into the topics we've covered in just the reverse order from the way we do here. First, uncertainty, then all the consequences. But I wanted to get you thinking first. I can summarize it this way: Nothing is certain, and less so the smaller it is. Isn't that unfair? Sounds like abuse of small things to me!

Uncertainty 47

You need to accept two things:

Quantum Law 1 from Werner Heisenberg (1927): for any object, the uncertainty of its energy and the uncertainty of its time of arrival vary in inverse proportion related by a very small number called Planck's constant, which is about 100 trillionths of a trillionth of a trillionth. Also, the uncertainty of momentum (mass times velocity) and of position vary inversely the same way—when one gets big, the other gets little—related by that same constant.

Quantum Law 2 says: if you measure something, you mark it.

There's no *reason* behind this. It just is. Einstein himself had trouble with that.

Practical consequences:

1. With big things, this doesn't matter at all. You'll never know the difference. If I want to know when my 1000 kg car arrived with an accuracy of 1 as (attosecond), its energy will be uncertain by 0.1 fJ (femtojoule). Who cares?
2. On the other hand, for something as small as a photon or electron, uncertainty becomes important. When I talk about light with one color, I can't really mean that. I can't have a pure color any more than I can have a pure tone. The uncertainty of a photon's color and its time of arrival vary in inverse proportion, governed by Planck's constant.

    This also applies to politics! The harder we hold on to "absolute truths," the more uncertain our position will be.
3. Another example: Like in Figs. 20 and 21, I put a pair of slits 1 cm apart 1 m downrange in a stream of helium-neon laser photons with a 1 kHz (1000 cycles per second) color uncertainty. This is really narrow compared to the laser frequency (color) of about 500 THz (500 trillion cycles per second). Now I turn on a detector that can tell which slit a photon came through. If I succeed, it won't interfere with itself in this setup. That means that the color uncertainty of that photon must have suddenly increased to 3 THz instead of 1 kHz, This corresponds to knowing where it was and when within 100 μm and 0.3 ps. Uncertainty says its color uncertainty had to increase because I measured it. ***Measurement changes things*** in the quantum world.
4. Light emission and lasers (Fig. 24): A hydrogen atom electron that is temporarily in the wrong orbit goes back home and emits photons with the wavelength shown in the figure. The energies of these are not arbitrary like in planetary orbits, but are just one quantum apart. One quantum is Planck's constant, 6.63E-34, times the frequency of the light emitted or absorbed in going between those orbits. Don't worry about the names in Fig. 24. They're named after the guys who saw black lines in the spectrum of the sun as the H atoms *absorbed* sunlight (see *Lasers*). Real lasers need more than just the ability to emit photons. They also need a sort of trap that makes an electron get stuck on its way home until another photon of the right wavelength comes along and "stimulates" it to get out of jail. Then, one photon becomes 2, then 4, then 8, 16 ... millions at the speed of light. This is *L*ight *A*mplification by *S*timulated *E*mission of *R*adiation, another

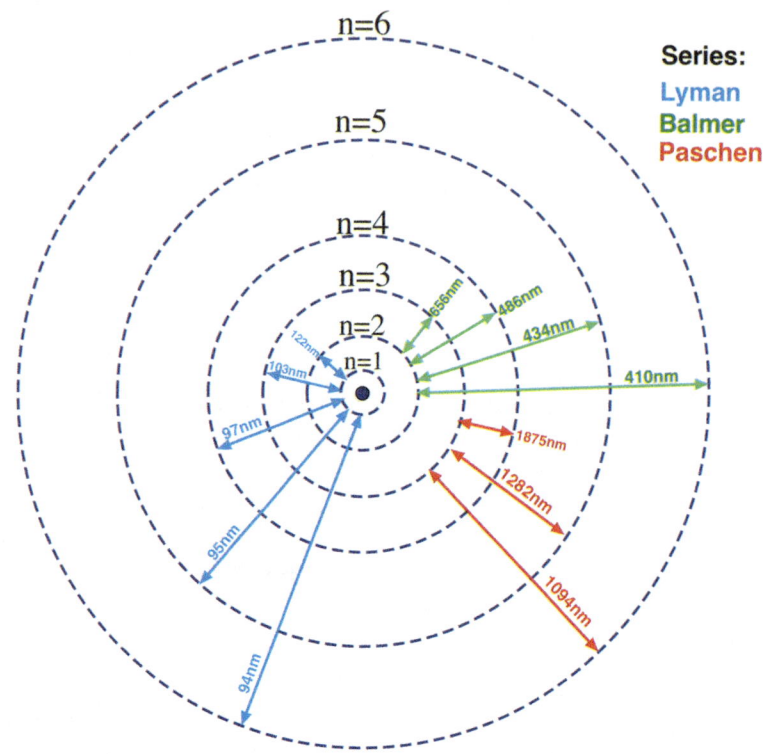

Fig. 24 Hydrogen atom energy levels and wavelengths (*colors*) of emitted light. 434 nm is *violet*; 656 nm is *red*, and 94 nm is way *ultraviolet*. H has just one electron (C. Phipps)

of Einstein's beautiful realizations, finally carried out by Ted Maiman in 1960. An old girlfriend of mine thought stimulated emission was a funny name for that. We'll look into it more in the chapter on *Lasers*.

5. A weird night-time experiment you can do that will convince you about photons: Camp out in the mountains where you can see the Milky Way and the stars around it clearly. Don't settle for the skies above LA. Look around between the constellations until you find a little, unfamiliar star that you can barely see, next to a brighter star that you can use for a sort of bookmark. Look away for a few seconds. Now, look back at the brighter star and stare at it. It will be a second or two before the little star pops into view! That's because the little star is about 2,200 light-years away, and it took that long for your retina to react to a light beam with a power of five photons per second (it takes five to activate a retinal cell)!

F. Wicke

Fig. 25 Heike Kammerlingh Onnes (Public domain)

Superconductivity

Any time you have an MRI, you're dealing with one practical application of quantum mechanics. Inside that electromagnet (where you go) is a 30,000 G field, 3 T (Tesla) in the metric system. That magnet would surely explode if you made it out of copper wire! You won't feel it at all, but the internal pressure trying to force the magnet windings apart is 36 times the pressure in your tires! And the electrical power it would take to make that field would heat any normal copper wire to the explosion point in a few seconds. So how do they do it? Superconductivity!

Normal wires have resistance. When you put current through them, they get hot. That's because the electrons keep bouncing off protons inside the wire. It takes voltage to make the current go past these obstructions, and voltage times current is power. If you turn it off, the current stops.

Back in 1911, Heike Kammerlingh Onnes (Fig. 25) at the University of Leiden discovered that, when it's cold enough, mercury becomes a superconductor. That means that you turn on the voltage once to get the current going, if you then connect

Fig. 26 Normal vs. Superconducting cables, both rated for 12,500 Amps. (Rama Wikimedia Commons)

the two ends of the wire together, the current goes on forever! In order to get it cold enough, he had to invent liquid helium: 4.2° above absolute zero, just 1.5° above the temperature of outer space! In his lab notebook, he wrote, "Mercury has passed into a new state, which on account of its extraordinary electrical properties may be called the superconductive state." He got the Nobel Prize for that. To dramatize his result, he carried his liquid helium thermos with the superconducting mercury in it by train and boat from Leiden to a Royal Society meeting in London, pulled the mercury out and let everyone watch the magnetic field suddenly disappear.

Practically, superconductivity lets you put an awful lot of current in a very small space (Fig. 26), and make things like MRI magnets.

Absolute zero? (See *metric system*), that's 273° below the freezing point of ice; in Fahrenheit, 459° below zero.

Since then, many metals have been made superconductors: aluminum, niobium, tin, indium … and also the temperature you have to reach has gone up a lot, but not yet to room temperature.

A fascinating fact: if you take a block of superconductor and bring it next to a magnet, it will automatically create its own surface currents that cancel the field and keep it from going inside. Ever.

There are only two hitches. It must stay cold, and there's a maximum magnetic field strength above which a super-conductor suddenly becomes a normal one, and whatever current is flowing suddenly turns into heat. For mercury, that was only 40 Gauss.

Fig. 27 B, C and S. (AIP Emilio Segrè Visual Archives, W F Meggers Gallery of Nobel Laureates)

Why am I mentioning all this *here*? As I said, it's a practical application of quantum mechanics.

It took until 1957 for theorists Bardeen, Cooper and Schrieffer to figure out how it happens, in their BCS Theory (Fig. 27). In other materials, it's still not so clear. It turns out that, to understand superconductivity, you *have* to think of the electrons as waves. They pal up as pairs that can sail through a low-temperature material like light. B, C and S also got the Nobel Prize for this work, in 1972.

Zero Point Energy (ZPE)

Look at that hydrogen atom again. *Why doesn't* the electron in that lowest state go ahead and emit one last photon and crash into the nucleus? We don't know exactly why, but it's not permitted. There is an absolute minimum energy for any wave, which is one half times Planck's constant times the wave's frequency. Trouble is, there are a lot of possible frequencies in all of space and if you add them all up you get something close to infinity. This is sometimes called the vacuum energy, and it might be the same thing as the dark energy I mentioned at the start of this chapter. Could you tap ZPE? People are still wondering about that, in 2015.

Black Holes

Imagine something so small but so heavy that even light can't escape its gravity! This is a black hole. We can't avoid this one, after the movie *Inter-stellar*. I think you understand the main idea, which has even been featured on *The Simpsons* in

Fig. 28 Sagittarius A*, the supermassive black hole at the center of our galaxy, seen here in X-ray light in a photo taken by the Chandra X-ray Observatory [*background* is an infrared light image]. (NASA public domain)

reduced detail. When a star dies, its core may collapse. It then has a huge mass in a small space and in some cases the gravity is so intense that it can crush the atoms and this thing just keeps collapsing. It can keep sucking other things in, crushing them too, gravity increasing all the time, until even light can't get out. Schwarzschild figured that out from Einstein's *general relativity* way back in 1916. It's amazing to me how much stuff was known before I was born, but was not part of a freshman physics course!

A black hole can be "super-massive," so big it can hold an entire galaxy together—like our Milky Way one (Fig. 28). The images in the figure were taken at three different X-ray "colors," the hottest being 30 k electron volts. The bright hot dot is *it*. What you see is SagA* consuming and heating gas to 180 million degrees, just before everything disappears.

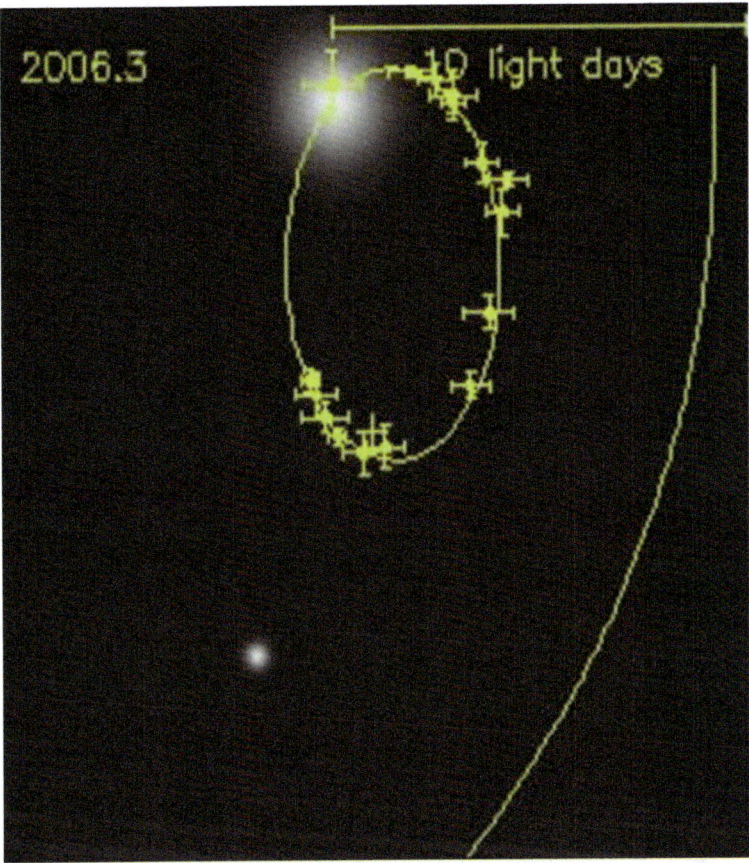

Fig. 29 The 5000 km/s orbit of SO-2 around the galactic black hole, data from the European Southern Observatory (Floppydog66 at en.wikipedia from Wikimedia Commons)

Get a star chart and find the constellation Sagittarius, down in the southern sky. In the northern hemisphere, that will be easier in summer. Over toward the right hand end of it is something called Sagittarius A* that you won't see, and couldn't unless you had an X-ray or a radio telescope. That thing is the black hole at the center of our galaxy. Sagittarius A* is 25,900 light-years from here (a mere 2.5E17 km, a quarter of a billion billion billion km!)

They know that, because a star called simply "SO2" orbits it every 15 years at 5000 km/s and a distance of 2E10 km (Fig. 29). From that they know SagA*'s mass is 4 million times our Sun's.

The galactic year (the time it takes our whole galaxy to rotate) is about 250 million years. By itself, the gravity of SagA* is not enough to account for that, and would give a galactic year of 33 billion years. But there are lots of other stars (240 billion of them—see *Numbers*) whose gravity also draws the galaxy together. You can think of SagA* as an organizing principle!

There are other black holes in our galaxy, but not all of them are supermassive. The one Lisa Simpson carried around on a stick was pretty small (at least initially!).

4D Ghost Stories

What about tunneling? If there are four *space* dimensions plus time, and if you're lucky enough that where you want to go is actually folded up close to us in 4D, then just walk through that porthole if you can find it. No reason it needs to take a lot of energy. Just a good 4D map! This is the plot of some of Philip Pullman's fascinating stories, such as "The Golden Compass," "The Subtle Knife," and "The Amber Spyglass." This is the wormhole idea.

What about 4D? What would that be? Unless you find one of those portholes, you can't experience it, so it would be pretty theoretical. Or just a nice thought experiment. String theorists use many dimensions, up to eleven, all the time. But those extra dimensions are curled up into little dots with a size like a billionth of a trillionth of a trillionth of a centimeter. Only three dimensions are supposed to be "unfolded." So what?

But I'm talking about imagining that reality has four unfolded space dimensions, only three of which we 3D beings can experience.

In 1884, an Englishman called Edwin Abbott wrote a fascinating book called *Flatland*. In the two dimensional world he designed, jails are little squares and the beings are triangles with more or less pointed heads. Now, imagine this: You live in 3D, and if Flatland existed, you could come into it *perpendicularly* and lift the prisoners out of their jail, or rob any bank. The poor Flatlanders would never know what happened. They would be unaware of you until your 2D *contact surface* (for example, your fingerprint) showed up in their world. Imagine bringing your finger very close to a pane of glass slowly. At first you make a dot in their world. Quickly, as you bring more pressure, the dot expands, having come from nowhere from their point of view. Then when you've done what you wanted, you disappear again just as strangely and there's not a thing they can do about it. They can't even understand what happened.

OK, think about one more space dimension. What if we live in "3Dland," but 4D beings exist? In the middle of your living room, or your bank vault, something appears from nothing. It starts as a dot and its intersection with our reality (a 3D thing) starts expanding. The door did not open, and the windows are closed. It grows as the pressure from the 4D Being increases. If you grabbed for it, it might move supernaturally quickly, because it's just an intersection, an abstraction. If the 4D Being tapped "Chopsticks" with its fingers on your world, objects could appear and disappear almost instantly on opposite sides of your room.

Fig. 30 A 3D demon preparing to cut your 2D world (F. Wicke)

Now, the Being does something in your world, *maybe to you*, and then disappears, again without going through any 3D walls. Think of ghosts, flying saucers abducting people, or of Christ suddenly appearing in the room with his Disciples after his death. Let's think about this by going down one dimension from our "3D" world.

Cutting World

Actually, what *can* that 4D Being do to you? Not much, you think, feeling safe, because you cannot move in that extra dimension, right? Or at least you think so. But—he can cut your world.

Imagine that you live on a loop of tape and that you have walked for days until you actually came back to your hometown again. Like going around the Earth, right? No problem (Fig. 30).

Now, the demon punches a hole in the part of the world where you live and carries it away, in another dimension that you have not experienced and cannot comprehend (Fig. 31).

If he wanted to … Now *you* live in a *Twilight Zone* world with an edge wrapped in a sort of fog all around that you cannot go past. It has atmosphere, stores, people, roads, and an edge just a couple of miles away that you somehow cannot travel past. Neighbors keep going out into the fog and coming back. Or not. Truly, now you can't go home again. We hope it's daytime.

Or, he cuts your world (Fig. 32). There's an *edge*! What *is that*? Where you used to walk, there is nothing, a Dead End. Suddenly, you can no longer go around the world.

Or, for amusement, he could bring the two ends of your film world together again but with a twist. Now, you live on a Möbius strip (Fig. 33)! As you again go around your world, when you reach the repaired edge, you suddenly find yourself in an alternate world that you never experienced, a parallel world just next to your own.

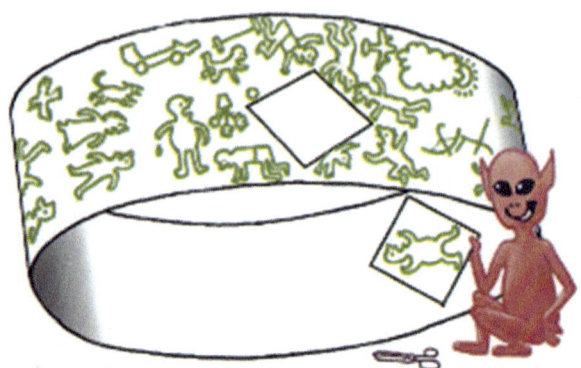

Fig. 31 Now he takes out a piece of your world and moves you in a dimension you cannot comprehend (F. Wicke)

Fig. 32 2D consternation! (F. Wicke)

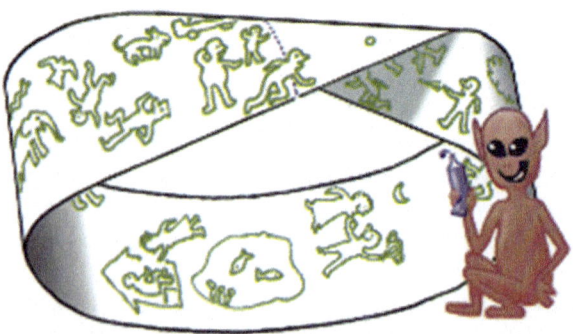

Fig. 33 OMG! A parallel world to explore! (F. Wicke)

In places, if the film where you used to live is transparent, you might see people you know from their bottom side, a side you never could see before. Like looking at someone *inside-out* in our 3D world, pretty disconcerting.

If he wanted to …

Conclusion

In science, many things are less well understood than you have been led to believe! Here, I have told you about a few of the things we do understand, some we only partly understand, what is at the very edge of understanding and what is still a mystery. Good luck as you sort these things out for us all in your science career!

The Metric System

Have you ever wondered why so many people prefer kilometers to miles and liters to quarts for measuring things, and about how their system was invented? This is the chapter for you!

Most of the world uses a way of measuring things called the metric system.

Some of the world still clings to familiar units like square feet [kind of a humorous idea in itself], inches, yards, miles, gallons, pounds, rods and furlongs. Not to mention degrees Fahrenheit.

Conversion among amounts of things in this so-called Imperial or English system is often inaccurate and tedious.

In contrast, the metric system is one of the most ingenious inventions of mankind! Why do I say that, and how did this system come about?

In the scientific, metric system everything is related to everything else by multiples of ten, from attoseconds (a millionth of a millionth of a millionth of a second: believe it or not, laser pulses are getting nearly this short!) to Exaseconds (a million, million, million), and even well beyond these limits. Scientists can use these numbers with some feeling for their relative size just by counting the zeroes. Those are called logarithms, exponents or powers of ten. A scientist may say "a light year is about 9.5 times ten to the 17 centimeters." [She means nearly an Exacentimeter!] Or that "an electron's mass is 9.1 times ten to the minus 31 kilograms," [technically, almost an hg (hellagram)!] and he will be able to have a pretty good feeling for what that means in his work. Why would anybody talk about, say, 100 cm (which is just 1 m)? As Prof. Low used to say at MIT, "it's because I like to think of things the size of my hand."

Scientists find the metric system essential. The world of science is one in which very large and very small things exist, and to count them by furlongs or feet would be disastrous.

How can numbers so big or small have any practical meaning for ordinary people? Our universe is about 0.4 Exaseconds old, for example, but you're not likely to live long enough to see a tiny fraction of that. Normally, the big and small numbers don't have much meaning for most people in their daily lives. But there's a very

Fig. 1 Inches and centimeters (C. Phipps)

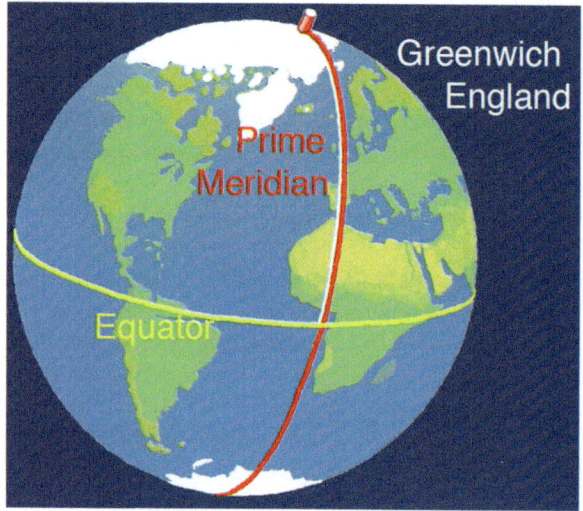

Fig. 2 The prime meridian (F. Wicke)

practical way units make a big difference. If I'm laying out a piece of wood to cut into four pieces, it's much easier to do that accurately using my tape measure with the millimeters and centimeters on it (Fig. 1), not the one where I have to divide 4 and 13/64 inches into quarters.

The French expedition of 1792–1798 measured the distance from the Equator to the pole, and called its length 10,000 km (Fig. 2)

Units are important. You may have read about how our Mars Climate Orbiter mission to Mars in 1999 was destroyed by units confusion. The engineers that built it used foot-pounds to describe the amount of momentum needed for final course corrections, while the physicists at the Jet Propulsion Laboratory assumed (surely!) that they used newton-meters, the metric unit. Whatever spreadsheets, tables or computer files were used to transmit this information between the two organizations failed to have units specified at the head of the column of numbers, and the misguided craft burned up in the Martian atmosphere as a direct result, also burning up 125 million taxpayer dollars with it.

So, how is the metric system ingenious? It's all about factors of ten, starting from a base unit that everybody agrees on. We have ten fingers, and that's still a good, human starting point.

Ten years after the outbreak of the French Revolution, tired of the arbitrariness of other systems, Lous XVI ordered the Marquis de Condorcet to establish a system of measurement that was to be "for all people for all time." In any language, the symbols were to be the same: km for the kilometer and A for amps, for example (named for the physicist Andrè-Marie Ampère, whom you can visit in the Cimetière de Montmartre in Paris). Incidentally, Tom Jefferson was part of the meeting in Paris at which these things were agreed in 1789. As we will see though, the metric system didn't make it in the USA.

It was decided that the basis of length would be the size of planet Earth, a pretty organic thing to do. The distance from the Equator to the pole needed to be measured, and two explorers were set to complete this task, on foot, horseback, and boat. They were Pierre Méchain and Jean-Baptiste Delambre, and they took more than 6 years to finish the job (1792–1798). The answer today is 10,019 km. Back in 1789 when the French established this unit, they measured 10,000 km. They can be forgiven for making an error of 0.2 % in measuring the distance! This is particularly true because the two surveyors were imprisoned several times and one died from yellow fever during the expedition.

Now that the world had a basic length, factors of ten above and below define myriameters, decimeters, centimeters, millimeters clear on down to the nanometers we use to measure the wavelength of light. There's a table to summarize those prefixes in the chapter on *Numbers*.

The length of a wave of red light is about 700 nm and blue light about 350 nm. Biological cells are a few micrometers (μm) in size, molecules about 110 pm (micro-micro meters!) or 0.1 nm (Fig. 3). Atoms are about the same size. So, light waves

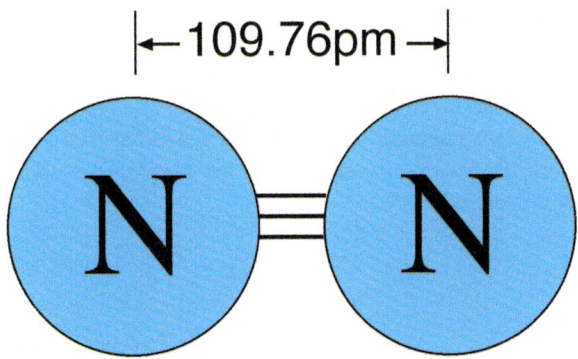

Fig. 3 A nitrogen molecule, composed of two nitrogen atoms (N) (C. Phipps)

are much, much larger. On the other extreme, our galaxy (we call it the Milky Way) is about 0.9 Zm (Zettameters, 100 million trillion meters) in diameter.

But that's nothing: our universe is about 0.8 Xm (Xennameters, billions of billions of billions of meters) in diameter. All expressible in powers of ten.

Exercise

There's no way I can show this range of distances in a picture. It would have so many powers of ten as to be meaningless visually. One way you can get a feeling for these numbers is the excellent book by Philip and Phyllis Morrison, "Powers of Ten." You turn a page for each power of ten, going past people on the beach to the whole universe! A way you can get a feeling for the true proportions in space is to draw a scale model of the solar system. I did that one day in the sixth grade, and it's fun. Get a long piece of chart paper or a roll of butcher paper, and tear off a piece 10 m long. Look up the distances from the sun and the diameters of all the planets, including poor, jilted Pluto. Now get out your calculator. Or if you're lucky enough to understand and use Excel, it will be a lot easier. I wish I had that in the sixth grade! Start at one end with the sun. We're making a scale model, right? No cheating. On this chart, 1 mm is 600,000 km! The largest object is the sun, and it's only 2.4 mm in diameter if you did it right. Not even the size of a pencil eraser! All the other planets are just specks, including Jupiter, about 1/4 mm in diameter! The first four planets all cluster around the sun within a distance a bit less than 40 cm. Pluto is that speck at the end of the chart. Stand back and look at it all. How can gravity hold that all together? How can this be called a "system"? How can puny people like us on that third speck send a spacecraft to the fourth speck? In Fig. 4, I've drawn the left-hand end of this diagram: the Sun and the first three planets to scale. Can you see them? I didn't think so. Try 500 %.

Based on the French expedition, for centuries after, the standard meter was defined as the length of an actual platinum bar kept in Paris, until about 1960. This length was transferred to other bars that were sent, for example, to the National Bureau of Standards in the USA. These copies agreed with each other to within 0.1 μm at the temperature of ice (zero degrees Celsius). That's one part in ten million.

As things progressed, the meter has been redefined time after time, because there are a lot of neatniks out there. In 1960, for example, it was redefined as "the length equal to 1,650,763.73 wavelengths in vacuum of the radiation corresponding to the transition between the levels $2p_{10}$ and $5d_5$ of the krypton 86 atom." Don't worry about it—it's in good hands!

| Sun | Mercury | Venus | Earth |

Fig. 4 The sun and the first three planets, to scale (C. Phipps)

Exercise

What else can we do with the meter? A pot that is 10 × 10 × 10 cm holds 1 L (liter). A cubic meter holds 1000 L. A liter of water has a *mass* of 1 kg (1000 g). Mass is the unit for the amount of stuff. *Weight* is something different, the force that it takes to hold up that mass against gravity, and it varies with where you are on the Earth, because gravity varies. The metric unit of force is the newton, named after Sir Isaac. From there we go on to millinewtons, micronewtons, and so on. That kilogram of mass weighs 9.8 newtons (they couldn't get that one to be exactly a factor of ten because of Nature).

As far as the English temperature scale goes, how and why did Dan Fahrenheit screw it all up in 1724 (Fig. 5)? He decided to start with the freezing point of *salt water*, naming that 0° F. Then he found that clear water freezes at 32 °F. Realizing

Fig. 5 Two temperature scales (Crop Service Center, Inc., Holland, KS. Used by permission)

Table 1 F, C, and K (scientific) temperature scales (C. Phipps)

Fahrenheit	Celsius	Kelvin	What happens
2.09E+08	1.16E+08	1.16E+08	Temperature in the atomic bomb
20430	11332	11605	Temperature of a one electron volt (1eV) plasma
9941	5505	5778	Surface of the sun
2800	1538	1811	Iron melts
1221	660	933	Aluminum melts
212	100	373	Water boils at sea level
193	89	362	Water boils in Leadville, CO
171	77	350	Nothing in particular
99	37	310	Your normal temperature
68	20	293	Normal room temperature
32	0	273	Water freezes at sea level
−109	−78	195	CO_2 freezes (dry ice)
−130	−90	183	Vostok, Antarctica, July 1983
−243	−153	120	Temperature on dark side of the Moon
−297	−183	90	Oxygen becomes a liquid
−320	−196	78	Nitrogen becomes a liquid
−423	−253	20	Hydrogen becomes a liquid
−452	−269	4	Helium becomes a liquid
−455	−270	2.7	Outer space temperature
−459	−273	0	As cold as it gets

that was a more fundamental point in the temperature universe he still chose to stay with zero for salt water, and decided that boiling water would be 180° hotter than freezing water, 212 °F, because the difference was *half the number of degrees in a circle*! Go figure. Anyway, this is how we in the USA wound up thinking our healthy normal temperature is 98.6, while everyone else thinks it's 37 (Table 1). Again, the C scale has replaced the F scale in almost all countries except ours.

And, yes, US engineers still use degrees Rankine, which are Fahrenheit plus 459.7.

Now what about metric temperature (Fig. 4)? The metric temperature range starts at the melting point of ice (0 °C) and ends where water boils (100 °C), again divided neatly into powers of ten. A pretty natural arrangement. What do scientists use for temperature? Well, they know there's nothing magic about ice as the lowest temperature you can find. The lowest possible temperature, absolute zero, is 459.7 °F or 273.2 °C below zero in those two scales. With absolute zero as the starting point, in what they call the Kelvin scale, your normal body temperature is 310 K, and water boils at 373 K because scientists just add the 273 degrees to the Celsius scale. Nothing can be colder than 0 K. Or can it? In 2013 one team claimed to have achieved a few negative nanokelvin. A nanokelvin is a billionth of a degree! This is the current record.

There's another important metric unit that doesn't have directly to do with length or volume: the energy unit, which is the joule. It is the kinetic energy (energy of motion) of 2 kg moving at 1 m/s, or about 100 g dropping a meter to hit your foot. In more familiar terms, the joule is a watt of power operating for 1 s. It's named after British physicist James Prescott Joule, who busied himself in the 1800s measuring how much heat was created while boring out a cannon barrel in water. The defense industry has always been a major source of scientific advances, unfortunately.

Another metric unit we will mention is the unit for pressure, the pascal. Atmospheric pressure is 101,325 of them. They didn't quite get that to work out evenly, again because of Nature, but it wasn't defined that way anyhow. A pascal is exactly the pressure of one newton of force on a square meter.

There are also what I call "specialist units"—for example, magnetic flux units of webers and electric charge units of coulombs (all named for famous guys), but you won't often run into them unless you're a physicist or engineer. Now that I think of it, you've probably heard of gauss, an older unit of magnetic field strength. A weber per square meter is a tesla and a tesla is 10,000 Gs. So there. The earth's magnetic field is about 0.7 Gs. One of the more humorous of these specialist units is the jerk, yes it is a metric unit, developed during the early atomic bomb era at Los Alamos. It is equal to a billion joules, 'cause they needed big numbers. You don't need to worry about them. But they're still all related to each other by factors of ten.

To understand why the metric system is ingenious, it helps to take a look at the so-called "Imperial" or English system of units for comparison.

In a sense, it *is* more human and personal than the metric one. But also completely arbitrary. Back in 1324, King Edward II of England used to have his foot measured when he left church on Sunday, and that length was the official foot! He also decreed that "three barleycorns, round and dry" laid end to end made an inch. Alternatively, the inch was the width of a man's thumb. Back then, the yard was the length of a man's belt (that could surely vary!) and the hand was the width of five fingers. Ah well, that's ancient history, you say? We still measure horses in hands. As late as 1959, the survey foot and the international foot were formally defined in the "International Yard and Pound Agreement," as if there's anything international about yards and pounds.

How are quantities related to each other in the English system of units? There are *twelve* inches to the foot, 5,280 feet to the mile, *three* feet to the yard, *sixteen* ounces to the pound, and the inch is subdivided by successive factors of *two* [1/4, 1/8, 1/16, 1/32, 1/64 inch]. The unit of land area is the acre [instead of the hectare, which is 10,000 m^2], and the acre is half a mile by a rod. A rod is 16-1/2 feet! The result is 43,560 sq ft to the acre. For irrigation, water is computed in acre-feet: the amount of water that would cover an acre to a depth of one foot. Try and turn that into gallons, let alone liters when computing the holding capacity of a lake! There are 640 acres in a square mile [or "section"] of land. Mass is measured in slugs and a slug under gravity weighs 32.174 pounds. Food energy units are Calories [kilocalories, 4,190 J]. The unit of heat is the BTU [British Thermal Unit, 1054 J] rather than the joule. Heat transfer is measured in BTU/square foot/hour, which amounts to 3.15 Watts per square meter. Thermal conductivity is BTU per square foot per inch per hour per degree Fahrenheit. A ton of air conditioning capacity is defined as the amount of heat required to freeze a ton of water! It's about 300 million joules.

No wonder the HVAC folks have such difficult educations! Oil is traded all over the world in barrels (42 gal, 159 L).

Amazingly in this modern age, ships (and jet airplanes crossing the Atlantic!) compute distance in nautical miles, which are 6080 ft, and speed in knots! A knot (1 nautical mile per hour) is 1.15 mph. The only reason for this is that there are about 60 nautical miles per degree of latitude, or 1 nmi per minute of angular measurement, these minutes being 1/60 degree.

In a few high-technology machine shops, people talk about "microinches," which are 25.4 nm.

I didn't even mention bars, buckets, bushels, candlepower, carats, cords, cubits, cups, degrees Baumé, degrees Brix, degrees Reaumur, dozens, drams, ells, firkins, fluid ounces, footlamberts, fortnights, gills, gradians, hogsheads, horsepower, hundredweights, kegs, leagues, long tons, lumens, magnitudes, nits, palms, pecks, phots, picas, pints, points, quarts, Röntgens, sacks, scores, shots, stilbs, stones, tablespoons, teaspoons, and troy ounces! Oh, and there's my favorite, bauds. You need to read Shakespeare a bit to see the humor in that. A kilobaud might be too much to expect.

But, before I make too much fun of the "English" units, these actually began in *Rome*!

No other modern nation beside the USA (and Liberia and Myanmar!—Fig. 6) uses these antiquated units in preference to the metric system. Go figure!

Of course, neither of these unit systems could do anything about the 365.25 days in a year, nor should they have. But it's a pity we couldn't have days divided into 10 somewhat longer "hours," each of these divided into 100 somewhat longer "minutes," and each of those into 100 somewhat shorter "seconds." (Fig. 7) The resulting second would be just 14 % shorter than the one we use now, and the minute 44 % longer. But you could tell 24-h time as 9:78:15 for example, instead of 11:28 *PM* and 34 s. I'm taking orders for a clock like that!

Fig. 6 The non-metric countries (Wikimedia Commons)

Fig. 7 Let's leave this party. It's 1:83:57! (F. Wicke)

Conclusion

The metric system is a genius invention. It allows us to think about big and small numbers, compare them and multiply them in a flash.

Exponentials and Instabilities

Have you ever wondered what people really mean when they say something (the national debt, the cost of Medicare, or whatever) is "growing exponentially" (or sometimes "geometrically")?

That's the subject of this chapter. If you deal with exponentials all the time in your work, you already know about these. Did you ever see the movie "Forbidden Planet?" When Morbius introduces the astronauts to the powerhouse on Altair 4, he points at gages on the wall, and says ominously: "ten times ten times ten times ten… almost literally to the power of infinity!" Well, that would be an infinite infinity, and we don't need that much. But you get the idea of exponentials, because each gage represents another power of ten.

Let's start with the old Indian proverb about grains of rice. The Raja wishes to reward the smart young lady Rani for something very good she has done for him. She asks the raja for only one grain of rice. He suggests she could ask for more, so she says "OK, send me twice as much each day as the day before, for thirty days!" In 30 days, the raja's storehouse is empty, because he has to deliver 27 t of rice on the last day and all told it's been 54 t. You can carry the example on: at the end of just 127 days, the pile of rice would exceed the mass of the whole earth (Fig. 1).

Another version of the story from 1260 AD features the wise Sissa ibn Dahir, who invented chess and showed it to King Shihram. The King was very pleased and asked Sissa what reward he would like. At first, Sissa answered that he didn't want a reward, but, when the king insisted, he asked for one grain of wheat on the first square, two grains on the second square and so on across the 64 squares on the board. Get out your calculator and tell me what 2^{64} is! You'll find it is 18.5 million trillion. In this story, the total is 2.6 trillion tons, and Sissa's request empties the storehouse of his king. We're not sure if bankrupting the king was a good move for either of them in the long run.

This kind of growth is called exponential, which just means that something grows without limit, and *the bigger it is, the faster it grows*.

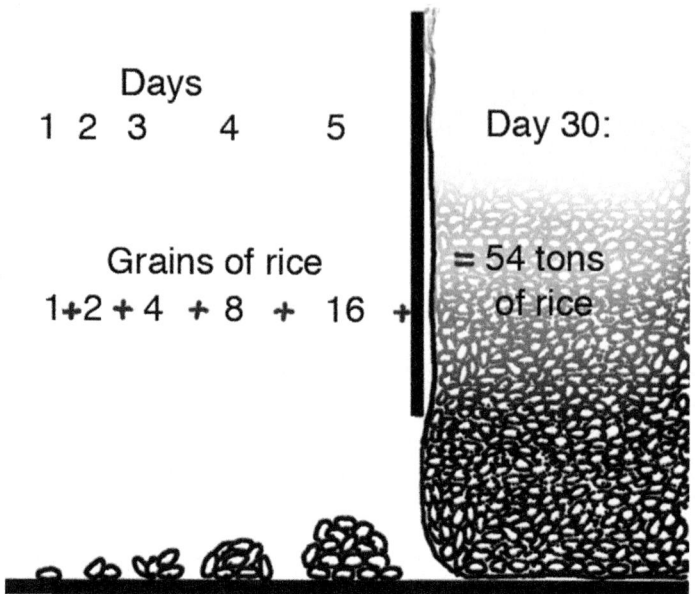

Fig. 1 Grains of rice (F. Wicke)

Table 1 Folding paper (C. Phipps)

Folding no.	Thickness (mm)	Size (cm)
0	0.05	100×100
1	0.1	50×100
2	0.2	50×50
3	0.4	25×50
4	0.8	25×25
5	1.6	12.5×25
6	3.2	12.5×12.5
7	6.4	6.2×12.5
8	13	6.2×6.2
9	26	3.1×6.2
10	51 (5.1 cm)	3.1×3.1

In the real world, this kind of growth can't happen forever. For example, the thing that's growing runs out of resources and dies. There's always a moderating influence that slows and stops exponential growth.

It's the reason why you can't fold a sheet of paper ten times. Try it. Get a large piece of very thin paper, say crepe paper, the kind you wrap fancy gifts in at Christmas. This paper is 0.050 mm thick (about 0.002 in.). Get a large piece, 100×100 cm. Fold it in half twice. It's now four times as thick (0.2 mm), and 50×50 cm. Now again twice: 0.8 mm thick and 25×25 cm. Four foldings, so far so good. But after ten foldings, you can see in the table below what you get. It is now almost twice as thick as it is wide (Table 1). While you can imagine doing that with a stack of postage-stamp-size cutouts, you won't be able to *fold* such a thing even ten times. It sounds easy, but use this to win some easy money from your friends!

In normal, natural growth, trees, people and animals quit growing when they reach the size they're supposed to be. When you reach that size, you're not supposed to grow any more. This is controlled by genes that regulate growth. When these genes mess up and lose control, we call it cancer.

We are often told that businesses and whole countries have to keep growing or die. The stated reason is usually that the number of people is growing, so we have more mouths to feed, or our competitor is growing. But why do these things have to grow? Why can't we just stop at a nice appropriate size like trees and animals?

The number of people doesn't have to grow, and the Earth couldn't stand it if it did. In fact, that number is *not* growing so much anymore. That makes sense. In the pond at our house, which is about 10 m across and 60 cm deep, there are 144 goldfish, plus or minus. That number has stayed the same for 5 years. Of course, they did that by eating their young, and "snakey" helped out, and so did some hawks. But anyway…

In the years since 2000, the US population growth rate has been a little less than 1 % per year. India's growth rate is slightly larger. China's growth rate has actually been less than that of the USA [contrary to what you've heard], about 1/2 %. Denmark has about 1/4 %, and Russia is negative at about $-1/2$ %. There was a long period when these rates were much larger, when resources seemed infinite. Now, it looks like the world is going to level off at 8 or 9 billion people.

If the number of people continued to grow at a 2 % rate like it was even up into the 1960s, in another 120 years, we'd have 100 billion people on the planet and life would be very difficult, people living in multistory coops like chickens, with inadequate food and water and huge waste problems. There are places in China that look like that already! Similarly, world energy consumption has grown at almost 2 % a year since 1820 to its present level of about 530 Exajoules (EJ). Exajoules are billions of billions of joules (see the chapter on *Numbers*); a joule is a watt of power working for a second. We talk about these units in the chapter on the *metric system*. To see where the limit is here, it's a fair estimate to choose 2 % of the energy that falls on the Earth from the sun each year. That's about 80,000 EJ. At the current rate of growth, we would get there in about 250 years. That sounds great, but only if we quit getting there by burning oil and coal, because we would see disastrous effects from global warming due to the carbon dioxide produced long before reaching that limit.

An annual growth rate of gross domestic product (GDP) of around 10 %, such as the Chinese economy has enjoyed in the last 15 years amounts to a factor of about 14,000 in just a century and can't be sustainable.

Instability

A close cousin of exponentials is instability. Instability describes a situation in which something starts growing without limit, all by itself. There are many kinds of instabilities and they are not all bad. A pencil standing on its point is unstable. The more it tips, the more it tips. A flag flapping in the breeze is unstable when it just begins to flap. Those are harmless ones. Financial instabilities happen when there's

Fig. 2 A feedback loop makes a squeal! (F. Wicke)

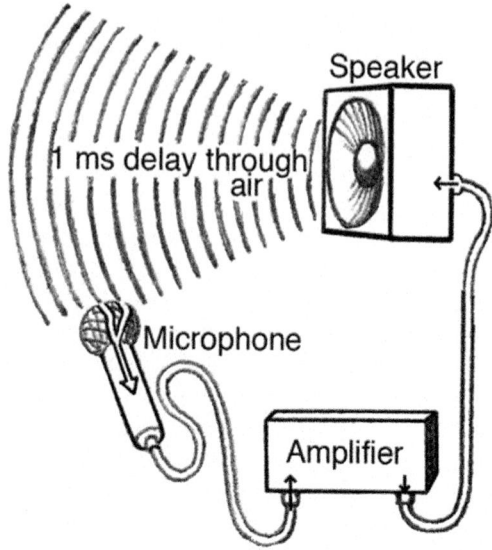

a "run on the bank" and everyone tries to take their money out at once amid growing panic, not necessarily produced by any real cause. That's a bad one.

Let's take a detour through an ordinary public address system to make this issue of "positive feedback" clearer (Fig. 2). You've seen it when someone sets up a PA system with the output volume of turned too high. In the Figure, "1 ms delay through air" just means it takes a thousandth of a second for a sound from the speaker to get to the microphone and go back through the amp.

Now you can imagine what happens. The amplifier takes the small current from the mike and makes it bigger according to the volume setting. The speaker makes a much louder sound which the mike picks up a short time later. Now the amplifier makes *that* sound bigger. This is a growing exponential. It only stops when the amplifier is putting out all it can, or the speaker blows. Now you hear that piercing screech that happens before you can find the audiovisual guy to turn down the sound, or you just point the mike away from the speaker, breaking the loop. The frequency will be higher when the speaker and microphone are closer to each other and lower when they're farther apart, as the delay changes. The other way to fix this is to put "gain control" into the amplifier design, so that the output volume turns itself down when things go "nonlinear". And you can also arrange for negative feedback. But while the system is in that state, even if you tried to speak through the microphone, the output of the system would still be a screech and people couldn't hear you.

People Instabilities

This is something like what happens when humans stop getting input and listen only to their own ideas, which is my main purpose in writing this chapter.

There are political instabilities, when each side of a conflict becomes more and more convinced that the only way to deal with their problem is to go to war, because the other side is lying, hates God, can't be trusted, and is interested in nothing more than killing us all, etc. Pretty soon, all you can hear or read in the news *on either side* confirms that kind of opinions.

When people used to live in villages with lots of relatives and neighbors nearby, if someone got to taking their own thoughts too seriously, the relatives and neighbors came over and straightened them out. These days, with each of us in our personal silos connected mostly by Facebook, Twitter, and the like, serious feedback based on personal knowledge of *us* that might correct the connection between our output and our thoughts (the input) are rare. So, for example, we get seemingly intractable political polarization because the people in each faction are not required by their culture or their living conditions to listen to the opinions of others. Some folks on weird religious trips deliberately fix it so they are walled off from other opinions and only listen to people in their group. Then you get a screech!

Another example: when you were young (or maybe you are still) did you think that some really old people seem to be like downspinning tops (Fig. 3), wobbling

Fig. 3 Downspinning top (Creative Commons Deed, Public Domain (Pixabay))

more and more as time goes on, until they spiral out of control? They seem to be getting more and more disconnected from the rest of us, more out of balance, but more and more certain that they're right, and less and less interested in your opinion. Do you have a relative like this? It's the same process. As we get older, fewer people come over to take a serious interest in *us*, challenge our opinions and offer theirs, and pretty soon our own thoughts are all that we hear. To match this, other people are less and less interested in our opinions.

Conclusions

Exponentials run entirely counter to our experience and judgment, so we have to do examples to show how they work. An instability is an example of something that can grow exponentially. Healthy things know when to quit growing.

Part II

Who *Really* Did It First?

Goals

In this section, I'm going to use a few key examples to illustrate scientific and technical knowledge in China, Greece, Rome, and the Islamic world over the past 3,000 years. This will not be a complete, detailed history of each culture, and will not list every single advance you may have heard of. We won't talk about why a triangle can be "isosceles," or how Archimedes made an infinite series that added up to the number π, or said "If I had a long enough lever I could move the World." Others have done very well on that already, and you can look up these things yourself. "Histories" in the linear sense ["In 1492, Columbus discovered America"] bore me as much as they do you.

I will use some surprising examples—it was not Columbus but the Greeks who first knew the earth was round, and they also figured out how far it is to the Sun and Moon before the time of Christ—to counteract the smugness which tends to creep into our Western European perspective. This is not so much to diminish that culture's achievements, which have been remarkable, but to give you a balanced point of view. We didn't invent or do everything first right here in America in the past 250 years! Or in Western Europe in the past 500. The first people to think of or build quite a few things were Chinese, or Islamic, scientists. One of these periods, about 500–1400 A.D., were the "Dark Ages" for us Western Europeans, who were wearing skins when science and literature were rapidly advancing in the Islamic world.

Also, I'm not going to go back to the beginning of our interglacial period and talk about the first guys to cultivate crops. That's another book! I have got to do things that interest me in order to interest you!

To repeat, our period is called "interglacial," and we some day may be glad for global warming! (Or not.)

If you read this section, I hope you too will feel you are a citizen of the world *as well as* of whatever country you call your own. A bit of xenophilia (BTW, both halves of that word have Greek roots!) to counteract xenophobia.

When I was in grade school, a graduation requirement was to demonstrate you knew one foreign language well enough to translate a scientific article. I took the easy way out with a language I had already studied (German). In those days, just 40 years ago, this was a practical requirement in science. It took 6 months to 1 year to get a Russian article in English, and *Russians* were doing a lot of important work—at least in plasma physics and lasers, my main thing. Glacial "machine language" translations supervised by native speakers were that slow to get an article from "JETP" to "Sov. Phys. Tech. Phys." in the Stanford library so I could read it. Now, Russians are still doing a lot of important work, but they all speak English.

Don't forget the dictum of Walt Disney (d'Isigny): "It's a small world, after all!"

Today, it is still true that knowing a language other than your own has practical applications! English is not completely universal yet. I ran into an example of this on a recent cruise down the Danube from Germany to Budapest. Our Russian captain told me that he needed two languages as we went further downstream to hear about local "road conditions" in real time from fellow captains on the river on his ship-to-ship radio. Indeed, at that point, all the noise on his radio was in Russian. If you want to pick up a girl/boy in Hungary, or France, it helps to know their languages rather than *expect* they know English, which is an insult. Asking for a date with the aid of Google Translator is not all that sexy. Not too long ago, I was in the queue for tickets to the upper levels at Notre Dame Cathedral in Paris just behind a vehemently English person who kept demanding the ticket agent speak English with never a word in French. With some amusement, I listened to her charge him two times what I paid a moment later. All it needed was "je voudrais un billet…" to save me 5 francs.

Does it take a lot of energy for you to imagine that history was the other way around and you had to learn Russian in order to keep up with the world? We English speakers are the beneficiaries of a worldwide event. Ours is the universal language, at the moment. There had to be one, and it probably had to be this one, now, for several reasons. Can you imagine control towers where controllers need to know French, German, Russian and Chinese in order to guide all the international flights?

Will English still be the *lingua franca* in a century? No telling. But—it wouldn't hurt to learn Mandarin if you're young!

Chinese Science and Art

You know the Chinese invented gunpowder, right? In 1044 A.D. (Fig. 1).

Gunpowder!

Now *there's* something interesting! May I digress? It was my 14th birthday, and I had gotten a present from Aunt Blanche. I worried a bit as I unwrapped it—she had a tendency to give me rubber tractors or books about traveling in the Holy Land that I would pretend to be pleased with before disposing.

But no! As I tore the last of the wrapping off I could see the words "Gilbert Chemistry Set" across a large flat box. A young scientist glowed with delight as he smiled at the purple stuff in his test tube, and I glowed looking at the box. Oh joy, oh rapture unforeseen! I glanced across the 20 small bottles of chemicals, found the instruction book and skimmed through it. Dozens of recipes on the order of how to make chlorine and bleach some cloth (meh); how to make invisible ink (look at that later); how to make a mixture that bubbled and fizzed (knew that already); decomposing water with an electric current (Hmmm!); how to make pyrotechnic "snakes" for 4th of July (Probably not as good as the ones you can buy).

Then, the jackpot: recipe number 47: "How to Make an Explosive Mixture." I disappeared into the backyard with the three important bottles, the instruction book and a match, while the adults droned on in the house.

"Take two Gilbert measures of potassium nitrate." OK! Gilbert measures were tiny things that might have been teaspoons for miniature dolls. "One of charcoal and one of sulfur." OK. "Mix thoroughly. Gunpowder is made from such a mixture. Stand back. Light." *Fssst*! It went up with a brilliant orange flash and a pungent cloud of sulfurous smoke, which smelled delicious to me.

I returned to the book. "Do not attempt to perform this experiment with proportions of chemicals larger than those listed above!" That sounded exciting! How *would* it be if we used teaspoons instead of Gilbert Measures? *Fooom*! All right! A toric cloud of smoke curled within itself like the cloud from an atomic bomb and sailed

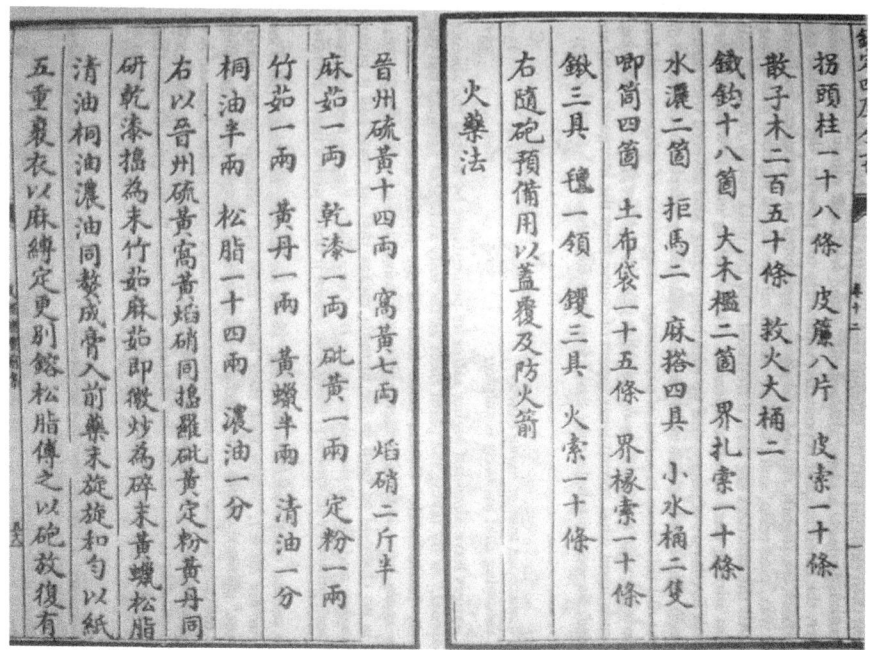

Fig. 1 The first written formula for gunpowder (Zeng Gongliang et al., *Wujing Zongyao*, a military manuscript from the Song Dynasty, 1044) By PericlesofAthens Public domain, via Wikimedia Commons). A Chinese friend tells me it says: Yan Xiao (potassium nitrate) 25 liang; Liu Huang 14 liang; Wo Huang (both, forms of sulfur) 7 liang—total sulfur 32 liang; Zhu Ru (dried bamboo, instead of charcoal) 1 liang. My friend says a liang is a unit of weight, at that time about 63 g. Assuming Ma Ru is also a source of carbon, the recipe is 42/2/32 % potassium nitrate/charcoal/sulfur, a very fuel-rich mixture compared to the optimum, least-ash recipe. The recipe also includes lots of glues and fillers

across the neighbor's yard before it disappeared, making their dog bark. Nobody saw it. But I had used up the supplies of those chemicals in the chemistry set.

All was not lost. In those days, many years ago, it was easy to find lots more of the ingredients in various drugstores in my town. If I used different stores, they'd never connect the dots! Potassium nitrate was a diuretic for cows, sulfur kept down chiggers, and so on. Next, the real problem was to determine the optimum ratio of the ingredients by weight, not just volume. Not only is this reaction one of those nasty three-component problems, but the products are carbon dioxide, potassium sulfide, potassium chloride, potassium oxide, and sulfur dioxide, in various temperature-dependent ratios, with reaction rates unknown to our city library—and there was no Internet then.

After weeks of late-night experiments, I solved the problem, repeating the process Mr. Zeng Gongliang must have gone through 1000 years ago.

What could the Chinese do with this invention? Well … amazing fireworks (Fig. 2) and firecrackers for one thing. They still do that. But, of course, even 1100

Fig. 2 Fireworks (S. Steele. Used by permission)

years ago, the scientific mind next turns to weaponry—because there's money in that—and my Chinese friend tells me Fig. 1 recipe also includes red lead oxide and arsenic trioxide, to make the smoke poisonous. Later, trebuchets, rockets, rocket bombs, land mines, hand grenades, exploding cannonballs, and the rest. But one *good thing* came out of this research: Man landing on the Moon! We'll continue this story as it affects *Rockets and Satellites* in that chapter.

Printing

Gutenberg invented printing in the early 1400s, right? Well, no. But, the blacksmith Gutenberg did invent movable type, right? Well, no. A gentleman called Bi Sheng did that about 400 years before Gutenberg, and the Chinese made the first printing press 900 years earlier than the German, in 593 A.D. Well, OK, the type was made out of clay instead of bronze or lead, and wasn't all that durable. But they did have excellent woodblock printing from 200, and used the technique to make whole books by the middle 700s A.D. Writing and printing are bootstraps for each generation of learners, an *exponential* process that advances all of civilization.

Chinese also perfected alcoholic beverages, even though the word "alcohol" is Arabic [see the chapter on *Islamic Science*].

And big bronze bells. Chinese metallurgists had already invented the blast furnace and the forge two to three centuries B.C. But, as far back as 1200 B.C, bronze technology was well developed (Fig. 3).

Fig. 3 Strange bronze head with lobster eyes from 1200 B.C. in the Sanxingdui Museum, Guanghan (Sichuan), China. (C. Phipps)

These figures apparently represent supernatural beings—of what sort, no one knows. My tour guide suggested they were from outer space!

Chinese developed steel in the 600s, and made sickles, swords, and sabers. And, of course, paper. The Chinese invented a crude paper a couple hundred years B.C., but a durable form was in wide use by 300 A.D. The first newspaper was printed in Beijing about 700.

By 200 B.C., China had 4000 miles of highways, and the one leading to Xi'an was made of brick.

Navies and Voyages

At least, Columbus and Magellan were the first to make huge ocean voyages, right? No. China was the world's leading sea power from the 1100s to the 1400s, with 20 squadrons of ships. A hundred years before Columbus, the great Chinese admiral Zheng He (also called Cheng Ho and Hajji Muhammad Shams) sailed all around the Indian Ocean as far as present-day Mombasa, and may have reached Australia in some accounts (Fig. 4). Born Ma He in a Muslim family, Zheng He rose from being a eunuch in the Beijing court to Admiral of the Seas. Zheng used huge ships, 130 m × 50 m, three times as long and wide as the Nina, Pinta, and Santa Maria. Can you imagine that career trajectory in the US Navy today?!

His largest treasure ships were perhaps 180 m in length (Fig. 5), and brought zebras, camels, ostriches and giraffes back to China. They had waterproof compartments, multiple decks, even swimming pools, and some could carry as many as 1000 passengers, according to Marco Polo. His largest fleet had 317 ships and 28,000 men. This included water tankers, supply ships, and horse carriers for amphibious landings, as well as teams of translators for the languages he was likely to encounter.

Navies and Voyages

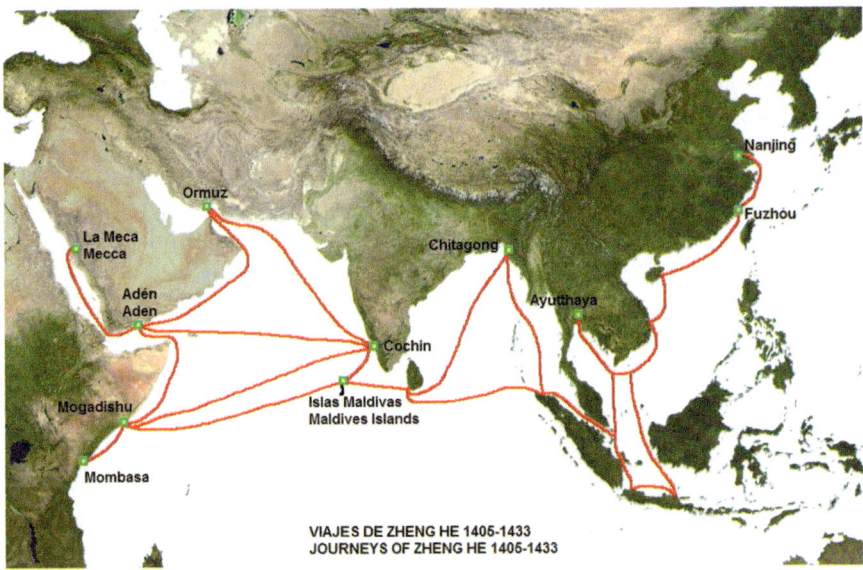

Fig. 4 The Voyages of Zheng He 1405-35. These were well-known Chinese trade routes established as early as 200 B.C. (By Continentalis, Creative Commons License, Wikimedia Commons)

Fig. 5 A treasure ship of Admiral He compared to the Santa Maria of Christopher Columbus [Photo of a display in the china Court of the Ibn Battuta Mall in Dubai] (Lars Ploughmann 2006, Creative Commons Attribution-ShareAlike 2.0 Generic license)

Admiral Zheng's base was Nanjing, the old capital of China, where his tomb is located (but not himself according to most accounts: he wanted to be buried at sea).

There is some evidence that Admiral Zheng's fleet sailed as far as the Americas, 70 years before Columbus, but that evidence is speculative, because as many records of his travels as could be found were burned back in 1435. Why this? In an historic and unfortunate reversal in 1435, the Ming Emperor ordered the Navy to be disbanded, and the ships and their plans and voyage records destroyed. He thought he had bigger worries (the Mongols in the North) than trying to rule the whole world. Further, while those foreigners could be seen to have interesting customs, he decided they could henceforth be ignored.

His Divine Highness believed foreigners were of no use to the most advanced civilization on Earth. They couldn't even speak Chinese! Also, they might infect the Chinese people with their barbaric practices. Admiral Zheng understood that it would be ineffective for his friends among the merchants to protest the Emperor's decision. The Emperor, like any educated Confucian, believed that merchants were a lower class of social parasites.

We will hear a similar story about the end of *Islamic Science*, although that happened because of religion. Curiously, these events happened at about the same time.

Silk!

Now I'd like to trace one Chinese technology from the Bronze Age to the present day.

From 3500 B.C. until a few hundred years ago, the most precious item traded across Europe for hundreds of years was silk. This trade made the Silk Road from China to Europe famous from about 200 B.C. With it, thanks to traveling salesmen like Marco Polo, Western Europeans learned about the Chinese, and they about us.

Did you ever wonder how they make that wonderful, silken fabric? It's not only miraculously soft, but durable (I've had a set of silk shirts for 15 years now) and takes a lot longer to show stains that require laundering. It ought to be incredibly expensive, and was in the past. But because of centuries of development in the manufacturing of silk, each of those shirts cost me about $30, custom-tailored.

The silkworm (Fig. 6) is a blind monster, selectively bred for 5000 years to do nothing but grow very big jaws, eat mulberry leaves, spin a cocoon (Fig. 7) and die. When it's done, it gets steamed to death in there.

Then, in a modern factory, a young worker puts it in a bowl of several others (Fig. 8), picks one strand of silk off the cocoon, and feeds that into a twisting multifiber strand of silk, one of thousands. These are then fed into a mechanical loom. Figure 9 shows a computerized loom in a Hangzhou factory 30 years ago.

Why am I taking you through all this? First, because I am fascinated by the ingenuity of human beings.

Second, silk manufacture is an example of humans involving all of Nature in a process, from finding out that *Bombyx mori* (fateful name!) is specifically attracted to the cisjasmone in mulberry, to using selective breeding to optimize the animal,

Silk! 83

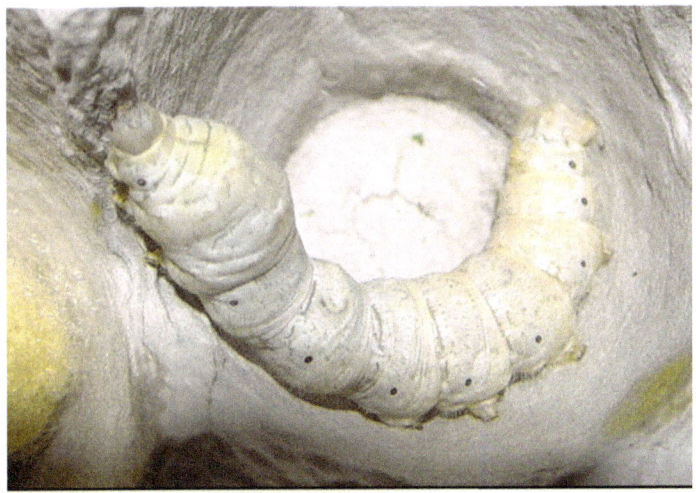

Fig. 6 A silkworm (Manyee Desandies. Used by written permission)

Fig. 7 A cocoon from the Hangzhou Silk Factory, 1987 (C. Phipps)

for a period almost as long as the development of modern crops or the domestication of kittycats. You could write another paragraph about how the parents were separately optimized to mate and produce monster children.

Third, silk manufacture is such a good example of the plus side of the industrial revolution: how human beings learn from past generations and dramatically improve complex processes over many years, and of how hundreds must cooperate, even today, in complex manufacture, in order to produce something that ordinary people can afford. Like a car, for example.

Can you imagine how much a tailored, multicolor silk robe, or a Prius, would cost if you and your friends made and assembled each part yourself, one at a time?

Fig. 8 In the Hangzhou Silk Factory, 1987 (C. Phipps)

Input Instructions Output

Fig. 9 Early computerized production: Hangzhou Silk Factory, 1987 (C. Phipps)

Conclusion

Chinese did a lot of stuff—from inventions to selective species breeding to exploring the world—that you may have thought Europeans did first. There's not enough room to talk about it all here, but you can go look it up. Or, if you'd like me to write a whole book about it, please send me an e-mail!

Next, let's look at science in Greece.

Space Geometry from Ancient Greece to Today

Geometry is easy, useful, and fun! I can tell you don't believe me. The way it's often taught, I can't blame you. But I will show you how powerful this tool is in this chapter, and maybe you'll find it interesting, too. You can measure the Universe with it!

Imagine it's 2200 years ago. You are one of two or three people on the planet that understands the Earth is round, rotates under the sun and that the Moon goes around the Earth, and you want to know how far they are from Earth. Everyone else thinks you're nuts, or sacrilegious. If you had to figure that out yourself armed only with pencil, paper, a protractor, and a long stick, your mind and one friend to help but without the Internet, phones, or books, *how would you do it*? Three Greeks named Eratosthenes, Aristarchus, and Hipparchos did it.

Have you heard that Columbus discovered the world is round? Did you know that statement is off by 1700 years? Or that Galileo was the first to figure out the Earth goes around the Sun? That's off by the same amount. They and their friends were just recovering from one of the greatest collective amnesias in human history, which we call the Dark Ages. In Western Europe, people had only recently been getting burned alive, or worse, for what they *thought*, and few people thought it was wise anyway to think about anything except heaven and the angels.

It could happen to us again if we choose dogma over curiosity, what somebody tells us is true over what we've found out ourselves. "And still, it moves!" Galileo muttered after lying to the Inquisitors to save his skin. That's why they call the age that started with Galileo the Enlightenment.

The Greeks didn't even have pencils when they figured out the local planetary geometry between 300 and 100 BC. But, they certainly knew the world is round, not flat, and they found out the size of the Earth and Moon, and how far those are from Earth.

In 240 B.C., Eratosthenes got the circumference of the earth right within 2 % and it is said he knew the distance to the sun within 1/2 %! He lived in today's Libya and was the librarian of the great library in Alexandria before it was burned, about 236 B.C., where he had the opportunity to study everything men knew at that moment. If you had *that* opportunity, would you do the same, or would you want to play Candy Crush Saga?

Today we compliment someone on being a "Renaissance man" or a "polymath" if they know and can do lots of things. Eratosthenes was a geographer, mathematician, poet, astronomer, librarian, music theorist, and athlete. He invented geography. His friends called him *Beta* because he was second best in the world *in every field*.

[See the chapter on the *Metric System*] His unit of measure was the stade—the length of a stadium—but what the heck. You gotta start somewhere! Why *not* a football field?! He invented the leap day, and made the first map of the world that used latitudes and longitudes.

Get this: not only did he know the earth is round, but he knew its axis tilts, and by how much.

Eratosthenes began by figuring out the circumference of the earth. A friend of his from Syene, now Aswan of Aswan Dam fame in Egypt, told him that at summer solstice the sun was directly overhead because its beams pointed straight down a well at noon. Eratosthenes already knew that Alexandria, where he lived, was 5000 stadia from Syene. On the solstice, he measured the angle of the sun to be 1/50th of a circle (we would say 7.2°), using a vertical stick. Knowing just those two things, he got about 39,690 km for his answer, just 2 % shy of the truth. You couldn't do any better anyway, using that technique.

How did he do that? Take a look at Fig. 1. He used geometry, and right here is a reason it's a good tool to have. Within a small angle (1/4°) light rays hitting the

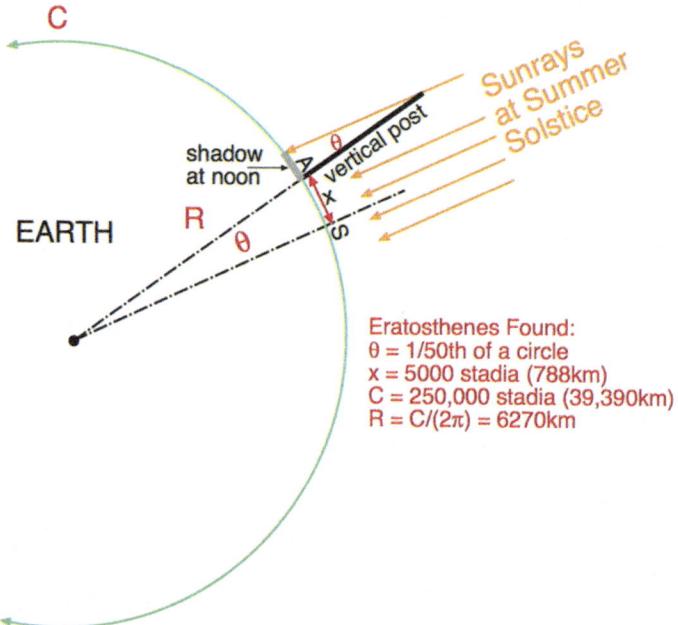

Fig. 1 A simple geometry problem applied to big stuff in 200 B.C. The right answers are: $C=40,074$ and $R=6378$ km (C. Phipps)

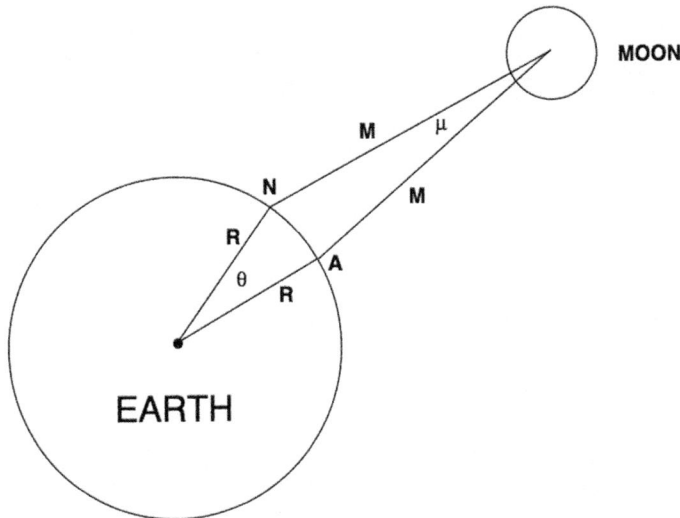

Fig. 2 (Not to scale!) A way to find the distance to the Moon, using parallax. N is Nicea and A is Alexandria. M is the distance to the Moon and θ is the angle between the latitude (43.6°) of Nicea and that of Alexandria (31°), about 12.6°. R is the radius of the Earth. Angle μ is the difference in the apparent position of the Moon from N and A at the same time. Trouble is, this parallax is too hard to measure without special instruments (C. Phipps)

Earth from the Sun are parallel, because it's so far away. I've shown them exactly parallel. So, knowing that the angle θ at the center of the Earth was the same as the angle θ of a sunray at A (Alexandria), and that it was just 1/50 of the whole circle, he could figure out the circumference of the Earth! C simply had to be 50 times the distance from A to S (788 km in our units). No tangents or sines here, just good clear thought!

Next, what is the distance to the Moon? Fig. 2 shows how one way it might have been done using parallax. You know how you can estimate the distance to a tree or a building by moving your head back and forth and seeing how far it appears to move? This is parallax. Because he already knew the radius of the Earth, he could calculate the Moon's distance from the angles.

The Greeks knew how to calculate where the Moon ought to be if you were looking at it from the center of the Earth (Fig. 2). A guy standing at Nicea (N) is not at the center, nor is one standing at Alexandria (A). Each one sees the moon in a slightly different place, and when you subtract the angles at which they see the Moon, you get the angle μ. Then, the ratio of M to R is the ratio of θ to μ.[1] This is parallax, the same rule we now use to find the distance to the near stars. But, it was too hard to measure μ, at that time, because it's only about 0.1°. Now, you and a friend could do it, using two Celestrons and an iPhone, in San Diego and Portland.

---

[1] Accurately speaking, the distance NA is $R\theta$, on the curved Earth, and about equal to $M\mu$ when all those angles are small.

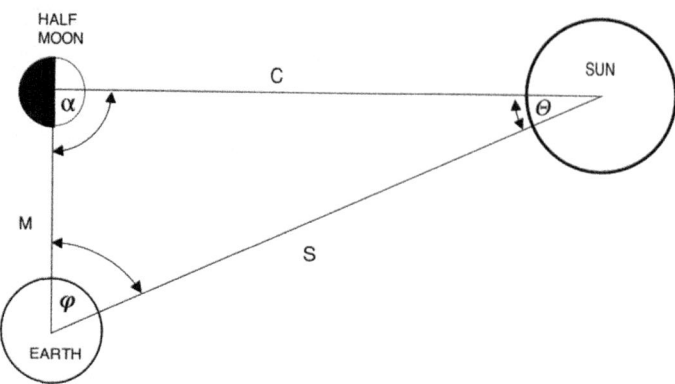

Fig. 3 A hard way to measure the distance to the Sun. If you know α and ϕ accurately, you can figure θ, because all three have to add to 90°. But θ is very small and it's hard to measure the other two angles accurately. Aristarchus measured ϕ to be about 87° and then estimated the Sun to be 18–20 times as far away as the Moon. Let's say 19, on the average (C. Phipps)

But the Greeks did measure the distance to both the Moon (M) and the Sun (S). How did they do that? The scientist Aristarchus (310–230 B.C.) is credited with getting the distance to the Sun (149 million kilometers) within 0.3 % 2250 years ago. A lot of books were destroyed when they burned the Library of Alexandria. People say "the original text was lost." Also, he was probably very lucky. Figure 3 shows how to get the ratio of S to M.

Aristarchus used the angle ϕ between the Moon and the sun when the Moon was exactly a half moon (Fig. 3). How did that help? Because then α has to be a right angle, 90°! It's a clever geometry problem, but hard to do accurately. If you can measure those angles at the same time exactly, then geometry will tell you the *ratio* between the earth–sun distance S, and M, the earth–moon distance. If you know M, you know S.

But he still needed to actually measure M in order to know S. Figure 4 shows how he used a total eclipse of the Moon do the whole thing.

The angle β is about 1/2°. This is the angular diameter of both the Sun and the Moon as viewed from Earth. You can figure that out yourself by looking at the Sun and Moon and measuring their angular diameter. Those angles are equal, because if the weren't we wouldn't have total eclipses of the Sun, where the Moon just fits over it.

Aristarchus found the distance S by watching a total eclipse of the *Moon* (Fig. 4). He noticed, and you probably have too, that the shadow of the Earth on the Moon in a moon eclipse is quite a bit larger than the Moon. It's sort of twilight on the Moon for a long time compared to the time of totality. So he was very clever. He guessed that the Earth's shadow during the eclipse was about two times larger than the Moon. Just look at Fig. 5 and you can see from the curvature of the Earth's shadow that it has to be quite a bit bigger than the Moon.

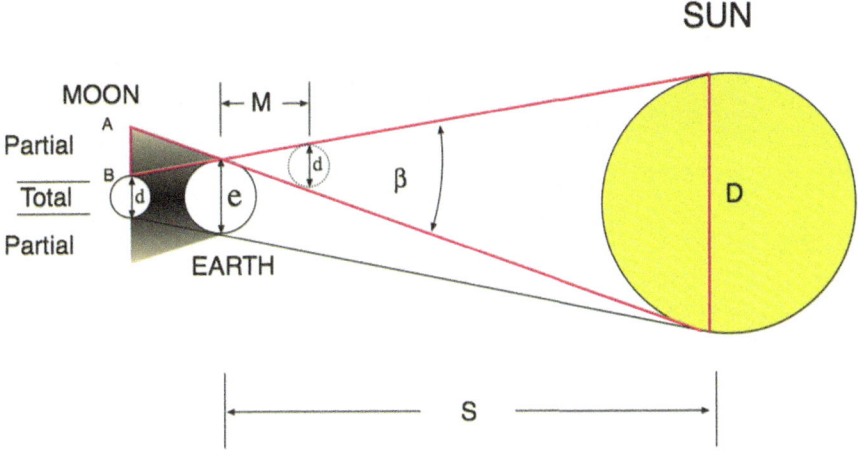

Fig. 4 How Aristarchus found the Sun's diameter about 250 B.C. He estimated the distance AB at about half an Earth diameter so the total shadow would be two times the Moon diameter. Actually, it's closer to one Earth diameter for a total of three. I've shown something sort of in between (C. Phipps)

Fig. 5 Moon toward the end of an eclipse (NASA Public Domain)

In Fig. 4, I've shown the Moon in two places, during the eclipse and a couple weeks later (dotted Moon). This way you can see that the ratio of D to d is the same as the ratio of S to M in that red triangle. If we think d is 1/2 e (this is R in Fig. 1, and we guessed that at 6270 km), and we *know* β is 1/2°, now we know M. Oh, and by the way, we know S and D! *Clever*, no? I think it's really neat!

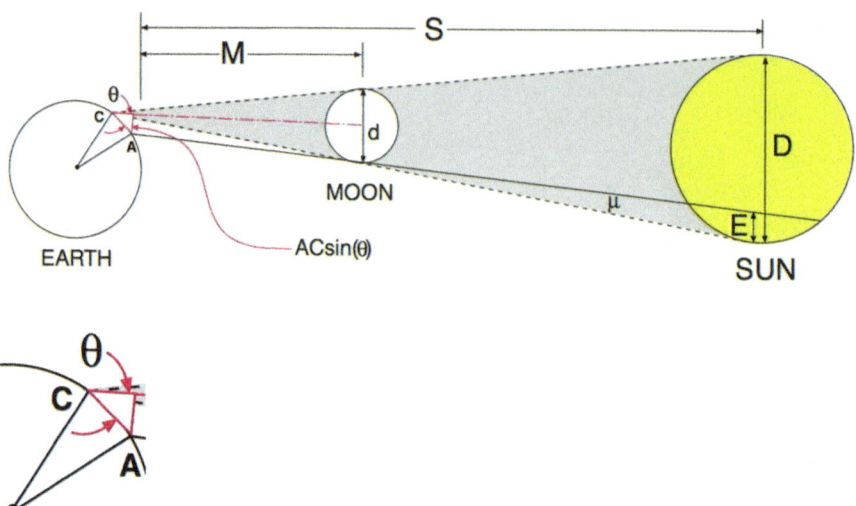

Fig. 6 (Not to scale!) A *hairy* way to find the Earth–Moon distance M from the parallax μ by the Greek scientist Hipparchus about 150 B.C. (C. Phipps). Inset: expanded view of the angles

Archimedes (287–212 B.C.) said that Aristarchus also knew that the Earth goes around the sun. However, like Galileo, he was threatened with an indictment for this belief so he kept it to himself. Even in ancient Greece! He certainly knew the stars were much, much farther away than the Sun because he could tell they didn't have parallax relative to each other during the year as the Earth goes around the Sun!

Finally, Hipparchus (190–120 B.C.) used an accident of history to find the parallax μ in Fig. 2 accurately. He's considered the founder of trigonometry, and lived in Turkey. Figure 6 shows his hairy but elegant solution.

He looked up the data for a total eclipse of the sun that happened before he was born. The story itself contained the answer about this small angle μ, without any further measurements. In Fig. 6, C is Çanakkale and A is Alexandria. As before, M is the distance to the Moon and S is the distance to the Sun. He knew his history. Once, when there was a total eclipse at Çanakkale, it was partial at Alexandria, and the people there saw $E/D =$ as 1/5 of a Sun.[2] At eclipse, both Moon and Sun are 0.545° wide, so μ is 1/5 of that, or 0.109°. I told you μ was small! For that particular eclipse, Moon and Sun were not overhead, but about 51° above the horizon [angle θ]. The distance from Alexandria to Çanakkale AC is 997 km in a north–south direction. Then the ratio $AC\sin(\theta)/M$ is equal to μ, and you get about 407,280 km for M. The real answer is 400,500 km, a little over 1 % error. Not too shabby, huh?

---

[2] Umm… This is a 1 % measurement, correct? It all depends on that "1/5 of a sun" being accurate to 1 % also. How did they do that? People have had lenses since at least 700 B.C. [see *Optics*]. But a simple pinhole can project an image of the sun as big as you like, and you can measure that very accurately.

My point is that these distances were known 2150 years ago, and Copernicus, Galileo and others just rediscovered them. Columbus was very far from the first one to know that the Earth is round.

Oh yeah. Did I mention *why* the Earth's shadow on the Moon looks so orange at the moment of totality (Figs. 4 and 5)? Lately, journalists have taken to calling it the "Blood Moon," as if all total eclipses didn't look the same. Imagine you're an astronaut on the Moon looking at the Earth during a total eclipse. What do you see? (Fig. 7).

Of course, no human being has ever seen this. That's because astronauts don't want to land and work in the dark. But, with *imagination*, a scientist knows this is what it looks like! If you're young, I hope you become an astronaut and live to see it!

Fig. 7 All the sunsets all around the Earth, at total lunar eclipse, viewed from the Moon. Sun is behind the Earth (C. Phipps)

Venus Transit: A Side Story that is Not Greek

Here's something a whole, fascinating novel has been written about [Thomas Pynchon's "Mason and Dixon,"[3] and it just amounts to a modern [1761] attempt to measure the Earth–Sun distance.

Here the principle is to measure the time it takes Venus to cross in front of the Sun from two different stations on Earth. Those are two different points of view thousands of miles apart, so Venus seems to go on paths a few thousand miles apart on the Sun and you get differences of a few minutes out of several hours in the times because of this parallax. Of course the sun is 864,000 miles across, so this difference is hard to see even today, just like it was in ancient Greece, but we have better instruments.

If we used two really good cameras, we could get a picture like Fig. 8, by overlaying the photos. Or, we can use the idea in Fig. 9, and figure out the angular distance between the two paths as viewed from Earth. Since we certainly know how

Fig. 8 Overlapping photos of Venus passing in front of the Sun from two locations gives you two dots close together (© 2012 Steven van Roode, used by written permission)

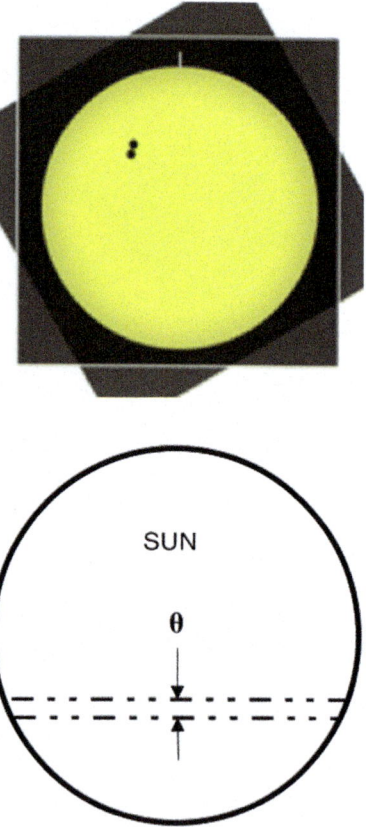

Fig. 9 Alternate way: distance between the paths seen at two locations gives angle θ, and you can get that from the time Venus takes to cross the circle (C. Phipps)

---

[3] Thomas Pynchon, *Mason and Dixon*, Macmillian 1997.

far apart those stations are on Earth, we have a complete triangle with two equal sides whose length is the distance from Earth to *Venus*.

Woops! We wanted the distance from Earth to the *Sun*. Ah, but we can cheat: *today* we know Kepler's third law, which says the ratio between Venus' year and Earth's year to the 2/3 power gives the ratio of the distances of those planets from the sun. The Earth–Venus distance V is just the difference between the Sun–Earth distance S and the Venus–Earth distance, $S–V$. From orbits, we know the ratio S/V. If we know both the difference and the ratio of two things, with a little bit of algebra we know the two things. You can work it out! In 1761, people (including our Mason and Dixon) went to the Cape of Good Hope, and to Siberia, Norway, Newfoundland and Madagascar, made those measurements and compared them. They didn't actually use cameras, but rather very careful measurements of the time the little dot took to cross the Sun, with an accuracy of seconds out of hours. From that they could get the lengths of the dotted lines in Fig. 9 and, because they're inscribed in a circle, using just the angular diameter of the Sun they could get θ by comparing the lengths.

All that depended on having the first really accurate *clocks* and Nevil Maskelyne's tables of longitude combined with accurate measurements of positions of the Moon at those times and longitudes.

Wow. This makes *my* head spin, as I'm sure it does yours. As I said, the result was good to within 1.5 % of the actual 149.6 million kilometers. Apparently, Aristarchus did better at 0.3 % [but he was *really* lucky]!

Of course (this is almost boring), today we know the distance to within 30 m. Takes all the fun out of it!

> Thro' our whole gazinglives, Venus has been a tiny Dot of Light, going through phases like the Moon, ever against the black face of Eternity. But on the day of this Transit, all shall suddenly reverse, as she is caught, dark, embodied, solid, against the face of the Sun, a Goddess descended from light to Matter...

And our Job, Dixon adds, "is to observe her as she transits the face of the Sun, and write down the Times as she comes and goes?"...

> Parallax. To an Observer up at the North Cape, the Track of the Planet, across the Sun, will appear much to the south of the same Track as observ'd from down here, at the Cape of Good Hope. The further apart the Obs North and South, that is, the better. It is the Angular Distance between, that we wish to know. One day, someone sitting in a room will succeed in reducing all the Observations, from all 'round the World, to a simple number of Seconds, and tenths of a Second, of Arc, and that will be the Parallax.
> Excerpts from Thomas Pynchon, *Mason and Dixon*, Macmillian 1997.

Conclusion

Before the time of Christ, Greek scientists and mathematicians did some heavy thinking that makes our minds spin today. They knew the Earth is round, that the Moon is a smaller object going around it, and the distance to it and the Sun using careful observation with primitive tools and geometry. One of them probably knew the Earth goes around the Sun, but was afraid to say so.

In the next chapter, we'll look at engineering accomplishments in the Roman Empire.

Rome

Concrete

Who first thought of liquid rock? And the Roman arch, and Roman roads all over Europe? Well, you guessed it! The Romans did 20 centuries ago, the finest engineers the world has known until now.

The very word is Latin: *concretus*, meaning compact or condensed, from *concrescare*, to grow together. Improving earlier recipes from the Greeks, Romans used concrete for 700 years from 300 BC, when they invented the first version stronger than mortar. Contrary to what you might think, Roman concrete had as much compressive strength as our Portland cement today. Structures from the Pantheon (Fig. 1) to the Colosseum to aqueducts depended on concrete. When Nero rebuilt Rome in 64 AD, his building code called for brick-faced concrete. It turns out that Roman concrete was better than ours in durability and compressive strength!

Roman engineers couldn't build suspension bridges and skyscrapers with it because they didn't know about rebar. But Roman concrete harbor embankments have lasted 2000 years, while modern Portland cement in seawater lasts about 50. The secret was lime and volcanic rock and ash. The "pozzolan" ash was an amorphous silica that enhanced the chemical reaction with lime to make a very strong material.

Roman Roads

Roman roads can be found all over Europe. In the same way as our interstate highway system was begun primarily as a way to permit missile transports to move quickly around the USA [you knew that, right?], Rome built these roads for rapid movement of their 180,000 legionnaires. Roman armies could easily move 40 km/day. Using a Pony Express-type service with post houses at intervals, the Emperor's messengers could travel up to 100 km/day and generals and VIPs in horse-drawn carriages could cover 120 km/day. Of course, in clear weather, signal stations could get a brief message across the whole continent in a day.

Fig. 1 Roman concrete 1900 years later: The Pantheon (125 A.D.), the largest unreinforced concrete dome in the world (Rich J. Heath, public domain)

Roman Text Messages

Digression: did you ever wonder how they did that? People say they used mirrors… that's false. They didn't know how to make such good mirrors. They had two sets of five flags or torches, and a code (Table 1). As you can see [*Numbers*], this is a base 5 system! Why not binary? If you're using flags, binary would require only one flag (up or down), but take five times as long to transmit a message. In this system, a single letter took two sets of flags or torches to transmit. "A" is 1flag, 1flag; "R" is 3flags, 4flags, and so on. They didn't use "Z," they didn't need "0" [no flag], and they hadn't figured out a base 10 system for numbers, so they didn't need "13." For that number, they would just use "XIII", right?

They had a lot of signal stations spaced just a few hundred meters apart. For example, Fig. 2 shows a map of all the forts on just a 15 km stretch in Scotland called the Gask Ridge. In spite of using 10 flags, they could only send 25 characters with the Table 1 system.

Think of Morse Code (Table 2). Do you see the similarity? But this is a base 3 system [nothing, dot, dash]. Morse could have taken much better advantage of a 5 flag system than did the Romans: only one set of 5 flags in sequence either up (−), horizontal (.) or down (blank) could have transmitted 243 different characters, instead of his 36, or the Romans' 25!

Anyway, on with the roads! Roman roads had foundations of clay, chalk and gravel, with larger flat stones on top (Fig. 3), sometimes surfaced with sand and

Table 1 Roman Signal Code

Flag1 No. ⇒ Flag2 No. ⇓	1	2	3	4	5
1	A	B	C	D	E
2	F	G	H	I	J
3	K	L	M	N	O
4	P	Q	R	S	T
5	U	V	W	X	Y

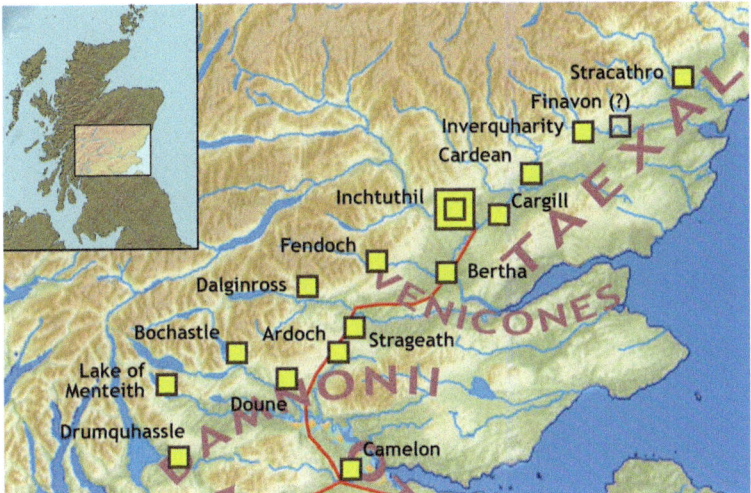

Fig. 2 Roman forts along the Gask ridge (Singinglemon Creative Commons License)

Table 2 Morse Code (1832) (Public domain)

A	•—	J	•———	S	•••	1	•————
B	—•••	K	—•—	T	—	2	••———
C	—•—•	L	•—••	U	••—	3	•••——
D	—••	M	——	V	•••—	4	••••—
E	•	N	—•	W	•——	5	•••••
F	••—•	O	———	X	—••—	6	—••••
G	——•	P	•——•	Y	—•——	7	——•••
H	••••	Q	——•—	Z	——••	8	———••
I	••	R	•—•	0	—————	9	————•

Fig. 3 A roman road in Pompeii (Paul Vlaar via Wikimedia Commons)

gravel. On boggy ground, bundles of sticks and other materials, or even pontoons, were the foundations. The roads were crowned for drainage, and often were literally "high"ways built above the surrounding countryside for better views and better drainage. Surveyors used theodolites and other modern tools. Main roads were 7 m wide, for six man columns to march, but sometimes as wide as 12 m.

Roman miles (milia) were about 92 % of ours, 1000 double paces. Makes you think there might be a connection between the words? There is.

While we're talking about dimensions, the benighted "English" units some of us are still plagued with are actually a matter of habit, unchanged in a major way since Rome! The Romans had feet (pes), square feet (pes quadratus), acres (one heredium = 1.24 acres), cubits, fingers, thumbs, palms and furlongs (one stadium = 0.92 furlongs)! These were practical units. A rod was the length of an ox-goad. A furlong was the distance an ox team could plow without resting. An acre was the amount of land one man could plow with an ox in 1 day.

Ultimately, there were 400,000 km of Roman roads, 50,000 km of them paved with stone. It's about the same ratio of ordinary to "super" highways we have today. Today, the US National Highway System has 260,000 km of roads, of which 77,000 km are in the Interstate Highway System. And you've probably read how the spacing of modern railroad tracks dates back to the spacing of Roman cartwheels.

Figure 4 shows a map of major Roman highways and roads.

Fig. 4 Roman road network, in Hadrian's time, 125 A.D. (Andrei Nacu, Wikimedia Commons)

"All Roads Lead to Rome"

Twenty-nine military highways radiated from the capital—hence "all roads lead to Rome." Napoleon did the same thing in Paris.

The soldiers themselves did the building, of forts as well as roads, a good way to keep in unusually great shape, ready for the next battle.

Where routes met, notice boards ("tria via," hence the modern word trivia) gave the news.

Pont du Gard

Even though they didn't have rebar, Roman engineers built even more massive structures than the Pantheon, which still stand today (Fig. 5), and that's better than the lifetime of any structures in the USA or Europe so far. Pont du Gard, near Avignon and Nîmes, is one example. Just look at that! Constructed about 50 AD from 6-t blocks almost without using mortar, it contains 50,000 t of limestone!

Fig. 5 Pont du Gard, built by Agrippa about 50 A.D. (Emanuelle, Creative Commons license)

The uppermost level is 275 m long. It cost 500 man-years of effort, using wonderful "modern" tools like the block and tackle [*Machines*] or windlass. Each day, the aqueduct on top carried 200,000 cubic meters of water [44 M gallons] for baths, fountains and homes in Nîmes. This is 72 % of the water supplied to the city of San Francisco today! The key to that bridge is the arch. Those arches were the secret for converting all the forces to compressive ones, which Roman engineers knew how to deal with.

The curved Roman arch was another technology that was lost during the Dark Ages and rediscovered later (Fig. 6). You can build one yourself, if it's small. Get a stack of concrete or adobe blocks and a piece of thick plywood. Cut the plywood in the shape of your arch, nail it to a 4×4, and brace the structure firmly so it won't fall over. I'm assuming the arch is to sit on a wall. Make some mortar and start building your arch, block by block from each side, with mortar in between to make each block act like a tapered one. The Romans did this by actually shaping blocks of stone with saws and chisels. The key item is that keystone block, which must fit exactly into its space. A day or so later when the mortar has set, remove the form and the arch will stay in place! That is, if the wall can take that sidewise force. I learned this by watching a local builder, Tony Sanchez, work on our home. When he finished something like this, he liked to say "That's purty!"

And then, for our final marvel, what else but the Colosseum (Fig. 7). It had 80 entrance gates leading to circular corridors from which staircases and passages

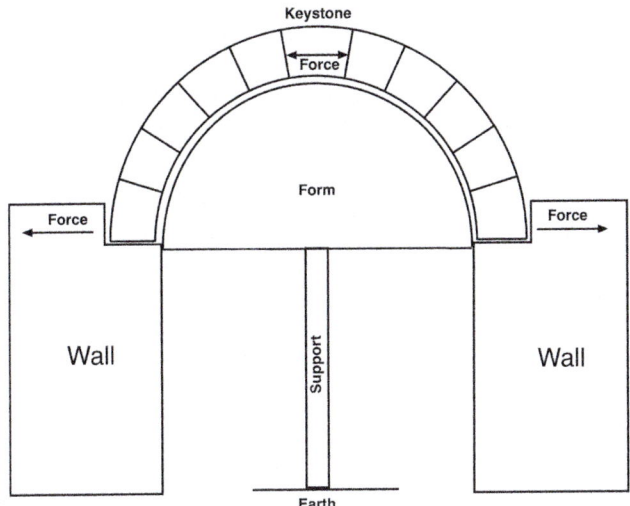

Fig. 6 Building an arch (C. Phipps)

Fig. 7 The Colosseum. Built by Vespasian about 70 A.D. (Diliff. Creative Commons license)

radiated to your particular seat. Those were called *vomitoria*, not because of Roman weightwatchers, but because this design allowed the stadium to *vomit* all of its 80,000 inhabitants within 10 min after the show! Today, we are accustomed to shuffling around for half an hour to get out of a hall built for 2000. The opera house in Berlin is the only modern structure I have seen that empties efficiently.

Oh. And did I mention safety pins (Fig. 8)? Here are two *fibulae* from about 100 AD. See the spring? Even the safety pin was not invented here—and in fact goes further back than Rome.

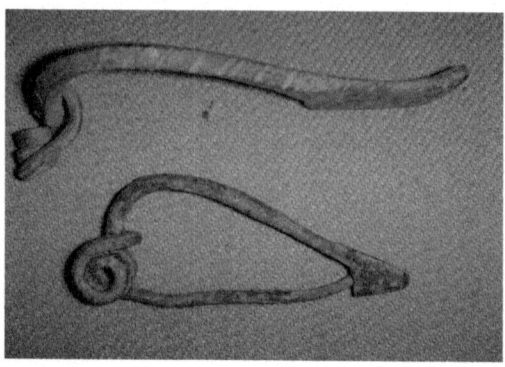

Fig. 8 Roman safety pins (Shawn Michael Caza, Creative Commons Attribution-ShareAlike 2.5 license)

We're all familiar with Roman art, mostly sculpture, which was very utilitarian and lifelike. Roman science and mathematics? Well, not particularly noteworthy. Roman theology and philosophy? Meh. One thing Romans did was to *write*, painting word pictures which allow us to know much more about their times than we would otherwise.

Plinius the elder was one of the best of these. His *Natural History* endeavored to summarize human knowledge up to his time, with chapters on astronomy, meteorology, geography, ethnography, anthropology, zoology, botany, drugs, medicine, magic, agriculture, metallurgy, sculpture and gems. I like the guy. He died in the explosion of Vesuvius trying to save his family, his friend Rectina, and the great library at Herculaneum.

My favorite Plinius quote:

> Men are most apt to believe what they least understand;
> and through the lust of human wit, obscure things are more easily credited.

You could say the same thing today.

Conclusion

The Romans developed magnificent processes for construction and transportation. Their engineers were unparalleled until modern times. The major structures that the Romans built have lasted thousands of years, far more than any that modern societies have put together.

Next, let's move forward to the Golden Age of Islamic Science.

Islamic Science and Art

What do the words Alcatraz, albatross, alchemy, alcohol, alcove, Aldebaran, alembic, algebra, algorithm, alizarin crimson, alkali, Altair, arabesque (Fig. 1), assassin, borax, café, canon, carat, carob, check, coffee, cotton, elixir, gazelle, genie, Gibraltar, giraffe, guitar, hashish, hummus, jasmine, julep, lacquer, mask, mile, minaret, monsoon, nadir, Rigel, rook, safari, sofa, sugar, tahini, tuna, Vega, vizier and zenith all have in common?

Why do we call the notes on the musical scale do, re, mi, fa, sol, la, ti? Do those names sound a bit like dal, ra, mim, fa, sad, lam, sin?

These are all Arabic words or derivatives. You can tell the ones that are direct from Arabic by the "al." Al-jebr means balancing, as in the two sides of an equation which, really, is all you do in algebra. Algorithm means a way of solving a problem, and comes from Algorithmus, which is how the name of the great Islamic mathematician Muhammad al-Kwarizmi got bastardized on its way into Latin (don't ask me how they got that!). He was the father of algebra. "Al" just says where the person was from. like the names von Engel, von Mises or da Vinci in the European world. Kwarizm is in Uzbekhistan.

Oh, and why are 1, 2, 3… "Arabic numerals"? To answer that, see *Numbers*.

There are ten centuries missing from most high school and some college courses on science, philosophy and art that we call the Dark Ages. They certainly were Dark for us western Europeans, but not at all for the scientists of the Islamic Caliphates, or for the Chinese, for that matter.

Caliphates? Islamic? OK, here's my problem: as a careful reader reminded me, I can't slide over the difference between Persians and Arabs, and call them all Arabs. The only thing they have in common is their religion. And, as much as I wanted to keep religion out of this, Islam is the fundamental common factor in the period I want to discuss. Islamic scientists don't even all speak Arabic, although most can. We don't have any Persian words in this chapter, so I won't make a big deal out of language.

Fig. 1 An arabesque ("Turquoise epigraphic ornament MBA Lyon A1969-333" Marie-Lan Nguyen, Wikimedia Commons)

Ibn Sahl

Let me start with this guy, because optics is my field, and I was so impressed when I learned about him a few years ago in a story from the Optical Society of America.

Ibn Sahl was an optical engineer in Baghdad. In 984 AD, he wrote a book titled *On Burning Instruments* that was the first serious mathematical study of lens design and optics. When I went to school they taught me that the Dutch scientist Willebrord Snell was first to figure out how a light ray bends when it goes from air to water, or another more refractive substance like glass. I'm here to tell you, Snell's Law is Sahl's Law! (Fig. 2).

I can't read Arabic either, but those who can say that little diagram in the upper left of Fig. 3 shows that the sine of the angle between the ray and the perpendicular to a surface times a property of the material we call the refractive index is a constant.

Light takes longer to go through glass, water and other materials than it does through space, and the refractive index is the ratio of the speed of light in space to that in the material. Ibn Sahl knew at least the practical results of that fact.

Fig. 2 Sahl's Law of refraction (Public Domain)

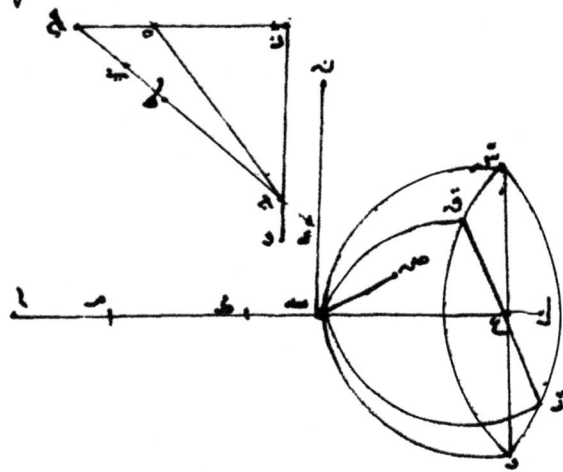

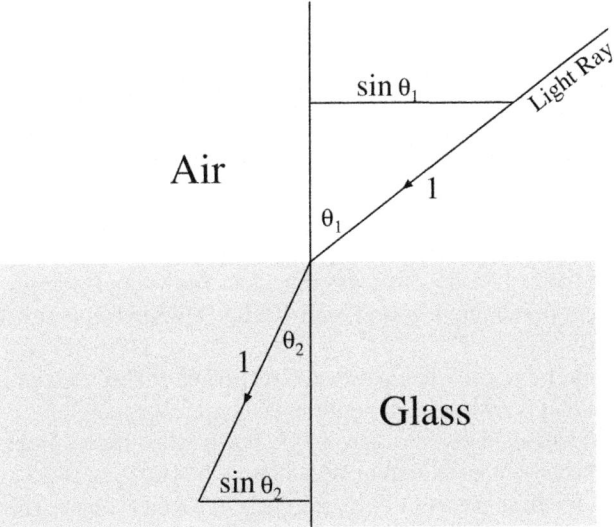

Fig. 3 Refractive index n is the ratio $\sin\theta_1/\sin\theta_2$ (C. Phipps)

Fig. 4 Sahl's aspheric biconvex lens (Public Domain)

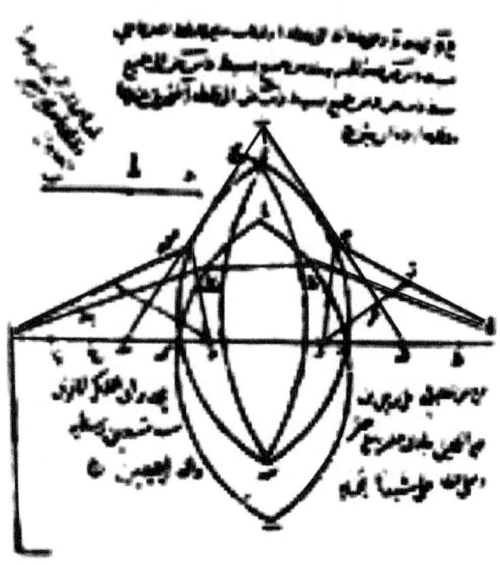

The lower right part of Fig. 3 shows a perfect, aberration-free aspheric (meaning non-spherical) plano convex lens design of Ibn Sahl's, whose convex surface is hyperbolic, a lot better than your eyeglasses.

Figure 4 shows his perfect biconvex lens design. He also understood and built parabolic and ellipsoidal mirrors.

Ibn Al Haytham

Acknowledged to be the father of modern science, ibn al Haytham (called al Hazen in Europe) wrote 14 books on optics alone. A true "Renaissance Man" a long time before the Renaissance, he was an optical physicist, astronomer, physician, chemist and mathematician. Born in Basra, he discovered the *camera obscura*, and knew from how that works that the image on our retina must be upside down. He figured out from double refraction in a water drop how rainbows work.

If you can believe it, the Greeks thought we see because our eyes emit rays. Haytham was the first to show that was ridiculous from simple logic. It seems obvious to you, dear reader, but it wasn't until 1021 AD, when he published his *Book of Optics*.

In that book, he states what is now called Fermat's Theorem, which says light takes the path that is easiest in going through an optical system.

Haytham understood that the reason light bends when it goes from one material to another is because it has different velocities in the two.

Figure 5 is the front page of this masterpiece. It's a busy scene. The naked guy is showing how standing in water distorts images. For some reason, the gentleman on the beach is using a concave mirror to project his image somewhere else. Mirrors on

Fig. 5 Front page of Haytham's *Book of Optics* (Public Domain)

the castle are setting fires on enemy ships—one has gone down and another is sinking. Not at all sure what the elephants are doing there, the rainbow is in the wrong place relative to the sun, and the guy's image should be upside down. I feel sorry for those poor slaves below decks manning the oars. You can see the author was thinking of weapons. From that time to this, defense brings contracts.

Ibn Musa al Kwarizmi

Kwarizmi is the inventor of algebra, and of the algorithm in mathematics. He lived a couple of centuries before the first two I have mentioned. In Baghdad at that time, the khalifs maintained a "House of Wisdom" in which he was a scholar.

He introduced Hindu numerals [see *Numbers*], knew how to solve quadratic equations, vastly improved the astrolabe and was the first Islamic geographer.

Astrolabe and Astronomy

The astrolabe! Now there's something! The astrolabe was a mechanical analog computer composed of several disks. It was a clock that operated day or night, a navigation assistant, and predicted the times of sunrise and sunset. The dagger-shaped pointers in the top disk (Fig. 6) show the location of major stars like Deneb or Aldebaran. The Greeks invented astrolabes in the 400s but the Islamic scientists refined them dramatically. Special plates for each latitude allowed them to be used all over the northern hemisphere, and they were still in use up until the 1600s in Europe. Chaucer owned one specialized for Oxford, and gave it to his son.

Want to build and use your own astrolabe? Check out these websites![1,2]

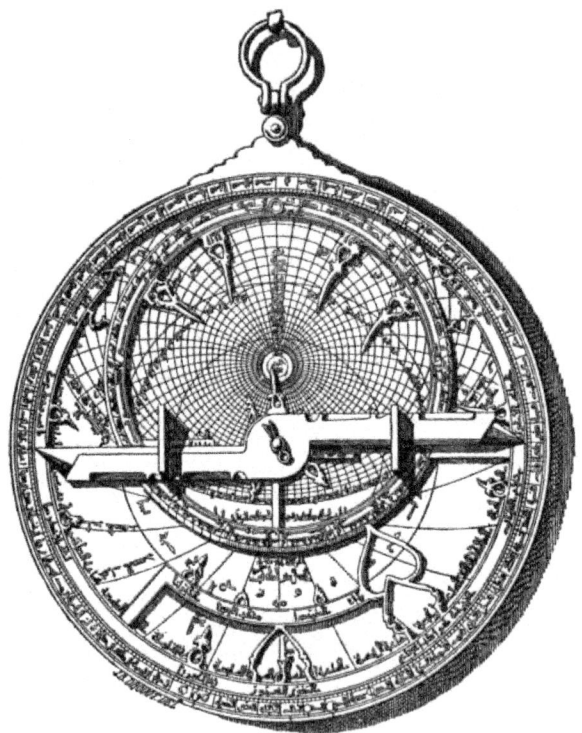

Fig. 6 Astrolabe (Public Domain)

---

[1] http://www.joh.cam.ac.uk/library/library_exhibitions/schoolresources/astrolabe/build.
[2] http://www.joh.cam.ac.uk/library/library_exhibitions/schoolresources/astrolabe/use.

The first astronomical observatories were Arab, built by Khaliph al-Ma'mun in Baghdad in the 800s. Some had instruments more than 20 m long for accurately sighting stars.

Al-Biruni

Abu RayhanMuhammad al-Biruni—born in a place called Birun, of course, in Uzbekistan in the late 900s, was a great mathematician and philosopher. He completed a textbook on spherical geometry in 1025, and developed techniques for solving cubic equations. I'll tell you, that's not easy! He knew how to find the cube root of a number. You know that a cube root of a number is another number which, taken times itself three times gives the number, right? Just like the square root of a number is the one that gives that number when you multiply it by itself. Al-Biruni knew how to take higher roots than cube roots. He developed a rudimentary calculus so he could describe the acceleration of heavenly bodies. He knew it was possible that the Earth rotated on its axis rather than the Sun going around it, but considered it an open question because data at that time didn't settle it.

Al-Biruni figured the circumference of the Earth from the height of a mountain and the angle to the horizon measured at its peak. His answer was 25,000 miles, within 1 % of the right value. Recall that [*Chapter 8*] Eratosthenes got within 2 % 1100 years earlier.

The thing I like best about him is that he said publicly that the Koran does not interfere with science, nor does it infringe on the realm of science.

Al-Razi

Born in Persia in the middle ninth century, al-Razi was one of the world's first great medical experts, and is the father of psychology and psychotherapy. Around 900, he wrote a book entitled *Why People Prefer Quacks and Charlatans for Skilled Physicians,* showing that not much has changed in 1100 years regarding the tension between traditional and alternative therapies. He wrote a 23-volume set medical textbooks titled *al-Kitab al Hawi,* containing the foundation of gynecology, obstetrics and ophthalmic surgery.

Omar Khayyam

You know him as a great poet, but this guy used a sundial, water clock and astrolabe to measure the length of the solar year (the time it takes for the Sun to come back to the same place) to within a fraction of a second in the late 1000s. A couple of my favorite Khayyam verses:

> Whether at Naishapur or Babylon,
> Whether the cup with sweet or bitter run,

The wine of life keeps oozing drop by drop,
The leaves of life keep falling one by one.

Ah, make the most of what we yet may spend,
Before we too into the dust descend;
Dust into dust, and under Dust to lie
Sans wine, sans song, sans singer, and—sans end!

Translating Scientific and Philosophical Works

Islamic scholars were particularly interested in learning from other cultures, and also translated what they could, especially Persian and Indian documents. In that period, Islamic scholars had a strong desire to gain the knowledge of the whole world.

Later, especially after 1400, at about the same time *China* turned inward, that thirst for knowledge came to an end, as religion came to be more important than science.

But, when Europeans were ready for it after the Dark Ages, during which religion dominated knowledge for its own sake, this information could be translated back into Italian, German or English.

Why did it take so long to transfer this knowledge? Up until the Renaissance and Enlightenment, to the European world, Arabs were Heathens. To some extent, that was true up until the 1800s when Walter Scott wrote novels like *The Talisman*, presenting Saladin (Salah-ad-din, born in Tikrit from a Kurdish family) as a mysterious, gracious person.

So then we have the European Renaissance and Enlightenment, and with it the genius Sir Isaac Newton.

Conclusions

We owe Islamic scholars a huge debt. For the most part, we wouldn't know what the Greeks thought if Arabic scholars hadn't dug it out and passed it on to us in the 900–1100 AD period. However, they added a huge amount of knowledge of their own, in optics, medicine, and other fields, to what we know today.

Now, let's jump all the way to Modern Science, a big jump indeed.

Part III
Modern Science and Engineering

Modern Science

Kepler and Newton

In the century leading up to 1670, scientific and technical knowledge on Earth suddenly turned a corner. Some nitpickers may disagree with me here, but I pick this time because it was the culmination of three events.

The first of these was the work of the Danish astronomer Tygge (Tycho) Brahe, published just prior to 1600. In Tycho, we have the true experimentalist, the type who believes in facts in themselves as reality, even when they contradict preconceptions.

Brahe's goal was simple: determine the position of the planets and stars as accurately as possible. He certainly wasn't the first. From *Greece* and *Islamic Science* you will remember that this goal goes way back. But he did that within 2 arc minutes, or 1/30 of a degree. This accuracy was revolutionary. For one thing, he found out that the time between the Spring Equinox and the Autumn Equinox is 7 days longer than the time from the Autumn one to the Spring one. What does that mean? The Earth can't be moving on a circle, can it? Equinoxes are two exactly equivalent moments for the Earth's orbit, and if it *were* a circle, it ought to take just the same time to go from Autumn to Spring as Spring to Autumn.

Did he use a telescope? No, even though his observatory, Uraniborg, cost 1 % of the Danish national budget to build during its construction. Just like the world of *Islam* 700 years earlier, he used a *quadrant* (Fig. 1), basically a very fancy sighting mechanism. But it was *carefully* built.

The second event: Brahe's assistant Johannes Kepler had the mathematical tools to make sense of Tycho's measurements, and he knew the planets go around the Sun. Although he wasn't the first to do that, he was one of the first to dare say it in public even in the 1600s. But it was easier to do then. You will remember from *Greece* that even Aristarchus had the same problem back in 250 BC, not to mention Galileo. Here he was relying on the axioms of Mikolaj Kopernik (Copernicus) from the mid 1500s who, in turn, had studied the work of the Islamic and Greek scientists. So nothing actually springs forth completely new, but these scientists were *geniuses*, especially Kepler.

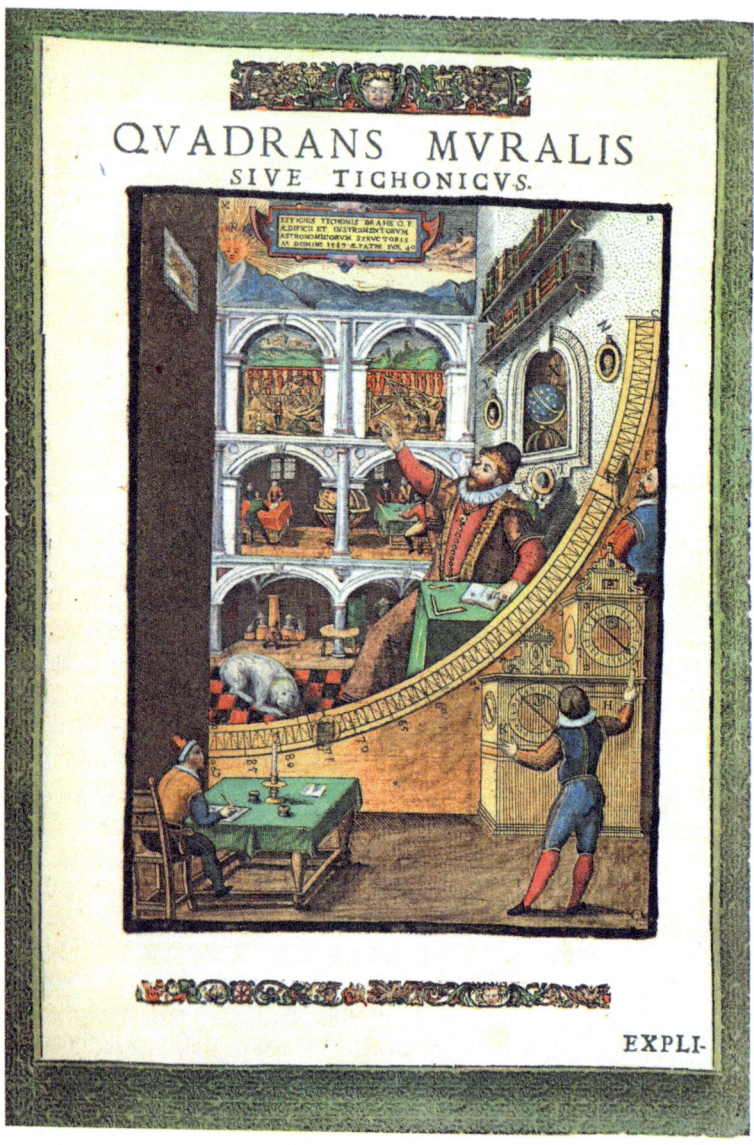

Fig. 1 Brahe's quadrant (Public domain via Wikimedia Commons)

Of course, he still carried with him searches for Divine Harmony, which led to weirdness like the *Mysterium Cosmographicum* (Fig. 2), which he believed revealed God's geometrical plan for the Universe. Each of the five *Platonic Solids* [not much has happened in a long time here!] could be uniquely inscribed and circumscribed by spheres which, nested together would make six layers corresponding to the six planets that were known at that time.

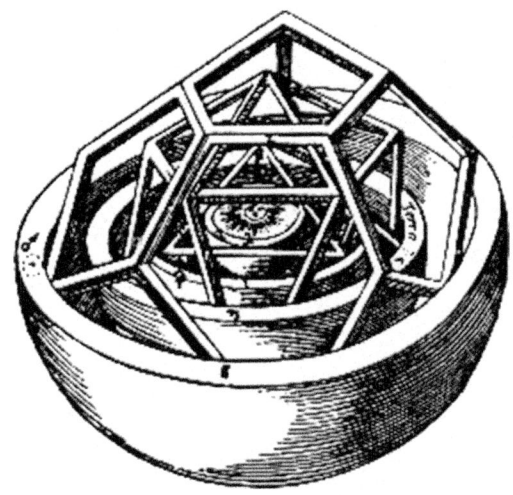

Fig. 2 Mysterium Cosmographicum. Mysterious indeed (Public domain via Wikimedia Commons)

Then, he saw a *new star*, the supernova of 1604. Contrary to Aristotle, *the heavens could change, after al!!*

Anyway, I have no desire to get bogged in historical detail here. The important thing is that because of those precise observations and the new framework to put them in, Kepler could dump the philosophical BS and rely on *data* to determine that planets must have *elliptical* orbits around the Sun (starting with Mars) and explain Brahe's observations. This was the beginning of modern science.

Elliptical orbits?!

Take a look at Fig. 3. The figure shows the orbits of two planets at different distances from the Sun. The sun is at one focus of both orbits. If you studied geometry, you know that an ellipse is a shape with two focal spots. A circle is just an ellipse where the two foci are the same. Every earlier astronomer thought orbits were circles and planets moved on spheres because the Heavens are perfect, and surely God, who lives there, would use the most perfect shape, right? *Sacrilege* to suggest otherwise!

He then applied his idea to explaining how fast Mars moves across the heavens, viewed from Earth, using Tycho's data. That's a *fiendishly difficult thing to figure out*, because *we* are planet 1, observing planet 2, and the rate at which it moves across our heavens is pretty complex. Sometimes you get *retrograde motion!* To start from that and say they're both moving on ellipses is … *genius*. Theory and trial. Trial and theory. It fit, within 2 arc minutes.

Kepler's four laws are the foundation of *modern* astronomy. Just now we talked about the first one. The second is that the Sun is at one focus of all orbits, not the center of a bunch of nesting spheres or Platonic Solids. By the way, that works only because the sun is 333,000 times as heavy as the Earth, and even a whole lot heavier than Jupiter, so it anchors the whole solar system.

His third law is that planets don't move at a constant speed but slower the farther they are from the Sun so that the *area* of those little triangles A1, A2 … in Fig. 3

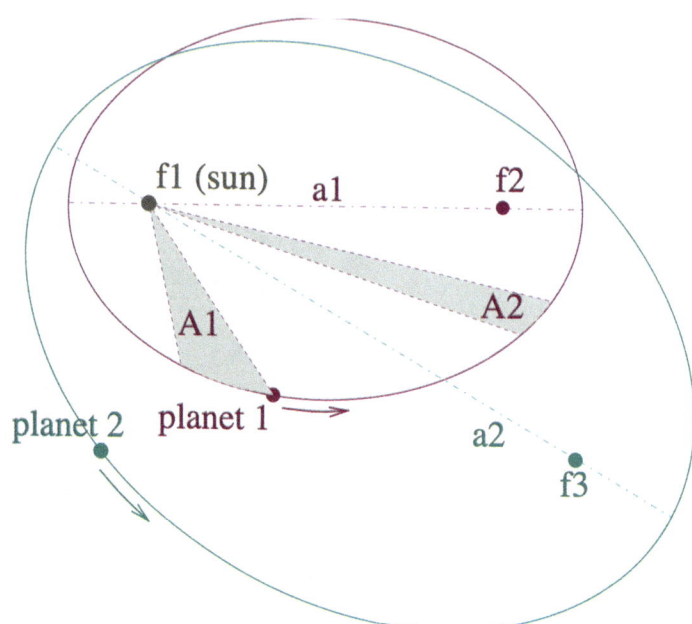

Fig. 3 Kepler's laws (Hankwang vis Wikimedia Commons)

stay the same as time goes on. A line between the Sun and the planet sweeps out the same area per second. Actually, Kepler thought the Sun was powering the whole thing, since something must drive it, right? Today we know that nothing needs to power an object in orbit unless there is friction to slow it down, like Earth's tenuous atmosphere (even at the altitude of some satellites). In the case of our manmade satellites, everything above about 700 km will be there for 10,000 years. Lower objects come in and burn up in a shorter time.

Now, the fascinating, unexpected, amazing fourth law of Kepler, the one we're most interested in here and which comes from the third one is: *the **square** of the length of the planet's year is proportional to the cube of the major axis of that ellipse*, the length of the dotted lines a_1 or a_2 in Fig. 3. Stated another way, a planet's year is proportional to $a^{1.5}$. And it does not depend on the mass of the planet, just that of the Sun.

In some way, that's as weird as the *Cosmographicum* in Fig. 2, *but it is supported by data*, not some *idea of celestial harmony*. It's true of the orbits of Earth and of the Space Station, and was true of Sputnik.

The *third event in our story*: Isaac Newton invented calculus, and figured out the force law for gravity. He looked like the scientific rock star he was (Fig. 4). Sad to say, I bet you already wonder if you will go further with this chapter because I said "calculus." Once more, this is because of the way it's often taught in high schools. Don't worry—we won't need calculus here, just show you what you can do with it.

Calculus is a mathematical way of computing the rate at which things change (differentials) and of knowing what the answer will be when you add up a bunch of

Fig. 4 Isaac Newton, the Original Geek (Sir Godfrey Kneller public domain)

Fig. 5 QR code for English Wikipedia. Can you read it? (Public domain)

small changes (integrals). Out of those ideas come *differential equations* like Maxwell's that we will talk about in *Electromagnetic Waves*, and integral equations. The latter can be used to get a spectrum out of a wave, or find out how to make a machine like your iPhone able to read one of those QR codes that are all over the place these days (Fig. 5).

Those Apple guys are cool, right? So are *Fourier Transforms*!

The philosophy that drove this invention and its coupling with precise measurements to produce modern science were truly singular developments in the evolution of man. Sorry, ladies, but that is a valid generic reference having nothing to do with sexism! In all previous history, people believed in the primacy of "pure reasoning," by itself, as a way of discovering how the Universe works. Well, guess what? You can think about Fig. 5 for the rest of your life and not be able to interpret it. It takes a machine with codes based on *integral equations*.

With Newton we have the theoretician who provided the mathematical basis of modern science.

Isaac Newton agreed that Kepler's fourth law *is* a strange result! Period is proportional to the *1.5 power* of distance? It must have to do with how gravity changes with distance, because mass doesn't. Kepler had already guessed that, BTW, but didn't have the imagination to go all the way with it to a precise, universal relationship.

Gravity holds the planets to the sun just like it holds us to the ground, and goes through the whole Universe! But *exactly* how does it change with distance from the Sun? The science students among you know the answer, but here's how we got it.

Newton knew that force is mass times acceleration. This was *his* first law. He also knew that if you fire a cannon shell horizontally over a cliff, how far it falls as time goes on is $gt^2/2$, where t is time and g is gravity. As we said in *Rockets and Satellites*, an orbit is just something falling *all the time* but going fast enough that centrifugal force keeps it up there (Fig. 6). How fast it falls does *not* depend on its mass, except for air resistance. Galileo knew that already in 1589 from his measurements at the leaning tower of Pisa, and he didn't need calculus to figure it out.

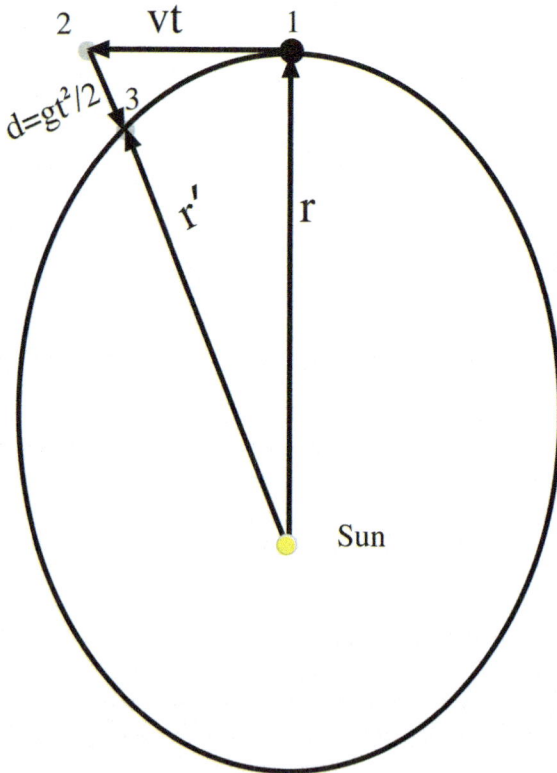

Fig. 6 Planet moving on an ellipse (C. Phipps)

Please look again at Fig. 6. The planet falls a distance $gt^2/2$ for short times where r' is about the same as r. On the other side of the ellipse, r' is certainly *not* the same as r, so there will be a different force on the planet, because it's closer to the Sun. For a circular orbit, centrifugal acceleration is the velocity squared divided by the radius of the circle. It's always balanced by *g*. It's what holds satellites up [see *Rockets and Satellites*]

Now for the other half of the problem. Here's where the falling apple story comes in! Of course, Newton wasn't hit by a falling apple, but he was out in the country because of a plague in London, and meditating on stuff (Fig. 7).

He cheated guessing the force law! It's always the best way in science, if it works. He didn't really solve the elliptical orbit problem and get the force law from that. Most planet orbits are pretty circular anyway. What he did do was to realize the Earth's gravity extends out to the Moon and infinitely far beyond. Then, he guessed gravity dropped off as some power of r [meaning r to the something, and what is that something?]. Then, he could find out how it falls off with r by comparing the acceleration of a falling apple on Earth (*G*) to the acceleration the Moon must feel to hold it in its orbit around the Earth (*g*). Even in the 1600s we had lots of precise experimental data about the Moon! For the apple, *G* is 9.8 m/s². Galileo already determined that number.

Fig. 7 Isaac Newton realizes the force law (F. Wicke)

Then Newton got out his pencil and paper. Because the Moon takes 27.3 days to go around the Earth at a distance of 384,403 km from Earth's center, its velocity has to be 1.02 km/s. For the Moon, g is v^2/r is 0.0027 m/s^2. That is 3600 times less than G for the apple on Earth. The ratio of the radius of its path from the Earth's center to that of the apple (6378 km *from Earth's center*) is 60. Sixty times the distance gives $(1/60)^2$ times the force. Guess what? The acceleration of gravity must go like $1/r^2$! As we move away from a mass M like the Earth, the acceleration of gravity is GM/r^2. Work out the details later, and give a paper at the Royal Society in London in 1687! When Newton started this project, *no one on Earth knew* if gravity goes like $1/r^2$, $1/r^3$, or maybe it's constant.

Why did I drag you through this?

To say this: using only measurements and math, not mysticism, it was possible for a scientist to guess that the force of gravity extends throughout the Universe, and to calculate how it varies with distance in order to explain at the same time the force of gravity on Earth and the lunar cycle, as well as Kepler's laws. Newton didn't have to measure the force! Turns out G is a pretty small number, 6.7E-11. Too small to measure anyway in Newton's time. Henry Cavendish finally did it in 1798.

Newton didn't have to appeal to the properties of heavenly things or pure thought by itself. Instead, he combined *his* pure thought with pure and accurate *measurement*.

Umm ... *why does it go like* $1/r^2$? The Universe wouldn't work if it didn't. But still ... what is gravity and why does it fall off like that? Nobody knows. See *Weird Reality*. Why don't you work on that for all of us?

Einstein

There are several misconceptions about our second genius to revolutionize modern science. Einstein (Fig. 8) was not kicked out of high school like you may have heard, but he did have to try twice to get into ETH Zürich because his French wasn't good enough the first time. What is true is that he couldn't find an academic job when he got out—none of his teachers liked him enough to write a recommendation letter! A Pisces, he had found university classes in 1900 to be too boring, and studied at home. But he did get a job as a patent clerk, and that gave him time to write down his "special" theory of relativity. We talked about that in *Weird Reality*, although we didn't say why it was "special." Another misconception is that he never earned a Ph.D. He did that in 1905. Having done more science than anyone since Newton by the age of 26, he was finally offered a teaching position at the University of Zürich in 1909, then at the Charles-Ferdinand University in Prague, then at ETH. By 1914 he was director of the Kaiser Wilhelm Institute for Physics, two years later was president of the German Physical Society, and had published his "general" theory of relativity. Not a bad career decade really!

The final misconception: his 1921 Nobel prize was for relativity. No, it was for the idea that light can behave as particles. That's because relativity was still considered too far-out.

Fig. 8 Albert Einstein in 1904 (by Lucien Chavan public domain)

Special Relativity

First, I'd like to get you thinking a little relativistically. Here's the problem in a nutshell.

In 1887, almost 20 years before Einstein published, the Michelson-Morley experiment we talk about in *EM Waves* got a lot of people thinking, especially Larmor, Lorentz, and Fitzgerald. The meaning of this experimental result was clear: the speed of light is the same whether you're moving or not. And yet if light were like sound, your motion matters. If you're a jet plane moving at 90 % of the speed of sound, the noise you make cannot go forward faster than an additional 10 %, because that's what sound speed is. In Fig. 9, that cone shaped thing at the back is trying to expand in the vertical direction at the speed of sound, *but the plane is moving at the speed of sound*, too. So, you get a 45° angle on both sides. Because the sound energy just piles up at the front until it tears the air apart, it can be pretty intense, and that causes sonic booms.

Now check out Fig. 10. You've had this experience, right? You hear a jet directly overhead, but you *see* it way out in front of where you heard it. It's because light goes a million times faster than sound. The symbol c_s indicates the speed of sound, 300 m/s. In the 40 s it took the sound to get to you, the plane already traveled 11 km. If it could honk its horn when you see it out there in front, the tone would be lower by 40 % because of the Doppler shift as it moves away from you. By the way, your eyes are not going bad. The plane in Fig. 10 is a bit blurry, because where it *really* is 1 cm ahead of where you see it!

Light did, after all, take 40 μs to come from it to you!

Fig. 9 Mach cone (by Realbigtaco GNU Creative Commons license via Wikimedia Commons)

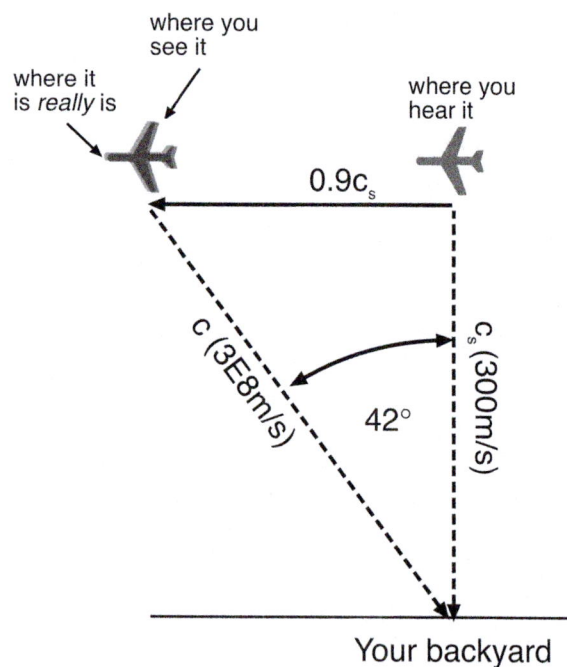

Fig. 10 Jet planes at 90 % of sound speed (C. Phipps)

Here is another thought experiment (those are the easiest kind!) (Fig. 11). A red pointer laser is flying toward you at 90 % of the speed of light. Is the red beam going at 1.9c? 0.1c? No. It has to be just c. But the color shifts! The number of wave crests between the laser and you has to stay the same whether you or a girl on the laser count them, or reality would be violated. She knows she is emitting red waves, but

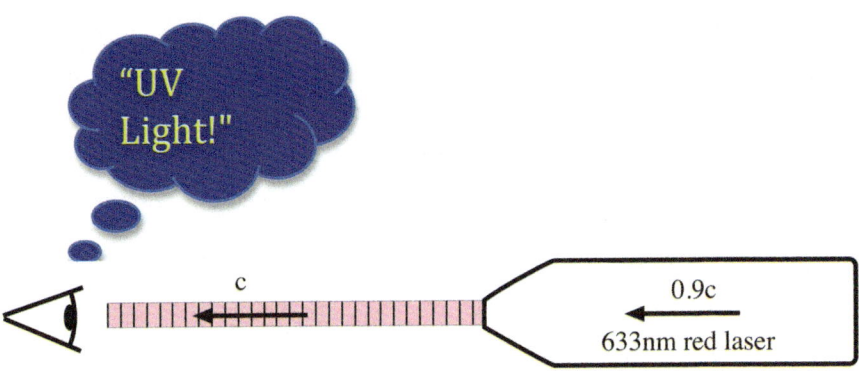

Fig. 11 Red laser going at 90 % of light speed (C. Phipps)

she *sees you coming toward her at 90 % of the speed of light,* scooping up 90 % of those waves every second. She guesses you see ten times as many per second as she is putting out. But she forgot the two of you measure time differently, by a factor of 2.3 due to relativity. The answer is that you see blue light with a wavelength 4.4 times shorter (10/2.3), or 145 nm, far into the ultraviolet. That makes my eyes glaze over, but it's the answer for relativistic doppler shift, at least on a straight line.

But I'm getting ahead of myself. What Einstein did was to see if he could make sense of a Universe in which that light goes at the same speed in any system, *and where physical laws are the same to everybody in their own system, whatever their speed relative to the others.* That's a tall order, but he did it! In the first part of this chapter, I ridiculed those who looked first for celestial harmony and tried to twist facts to fit their ideas. Perhaps I was unfair. That urge has driven all great science. The difference is that in this century we can measure stuff, very accurately and test postulates like Einstein's, and violating some religious precept is not an important consideration in science. So far as anyone knows a century later, he was right.

In order to make it all work when I, sitting here, look into another system moving at velocity v, not only my idea of time and distance in that system have to change with the ratio v/c [Lorentz and Fitzgerald knew that], but electric fields, magnetic fields, forces, masses, accelerations, the idea of simultaneity—the whole works has got to change, so that Maxwell's equations still work with the same value of c. [*Electromagnetic Waves*]. My alter ego *in that other system* sees nothing out of the ordinary! It does get complicated and we will just talk about some of the amazing consequences in the rest of this chapter. Einstein's was a beautiful dream, and he succeeded in realizing it in 1905. The trick was to make spacetime, the square root of $[x^2 - (ct)^2]$ the same in all systems.

Imagine the jet in Fig. 10 is going overhead at 90 % of the speed of light! The plane's actually out there 42° ahead of where you saw him. So far, so good (Fig. 12).

But, a bunch of weird things happen. The pilot doesn't have a horn, but he does have a red laser and has been shining it straight down all during his flight, just for fun, although he knows he really shouldn't do that. When you see his beam, it will

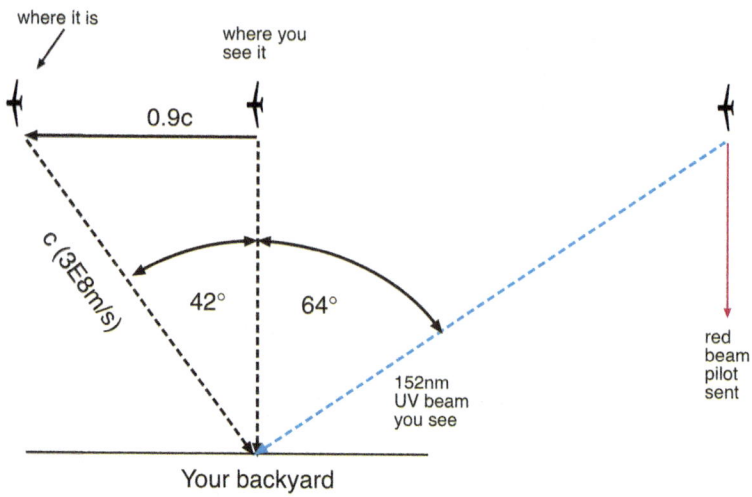

Fig. 12 Jet planes at 90 % of light speed (C. Phipps)

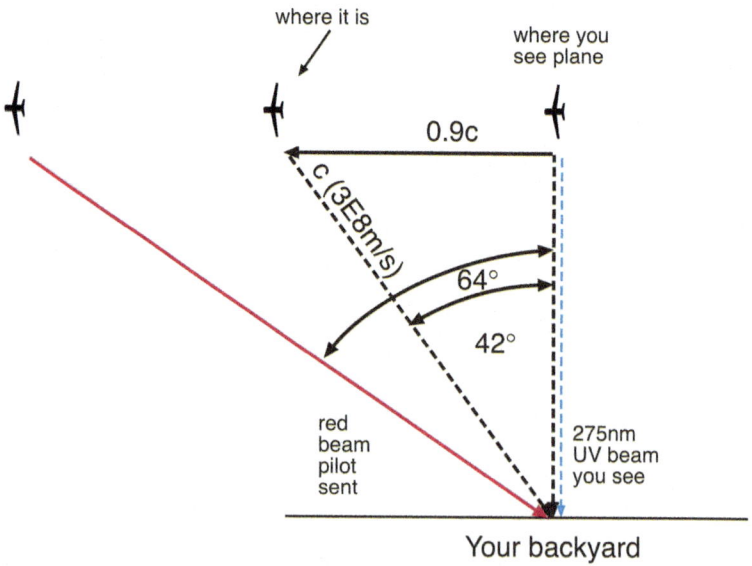

Fig. 13 Jet planes at 90 % of light speed, a bit later (C. Phipps)

be 152 nm in the hard ultraviolet. It won't get through the atmosphere anyway but if it did I wouldn't recommend you look at it.

A bit later, an even stranger situation (Fig. 13). By happenstance, the pilot also decided to shine his red beam backward instead of at the ground, 64° from vertical. If you're looking up, you see it coming straight down, but at 275 nm in the ultraviolet! There is a *relativistic* doppler shift, even when something is going straight past, not at you! This does not happen with sound, and is different from what you expect. At least, the color you see is a bit redder now that he's going *away* from you instead of toward you, but it's still UV!

Oh, ah ... notice anything strange about the planes in both figures? Yes, they have shrunk tremendously in the direction they're flying.

For the pilot, things are weirder yet. Where before there were stars kind of uniformly all around, now the whole universe seems to have turned red and blown back behind him, except for directly ahead, which is blinding ultraviolet. Also, since he is the one flying, not you, he's happy that you're aging 130 % faster than he is. We talked about time shifts in *Weird Reality*.

But: it's not all relative! The equations are symmetric, but lingering effects are the consequence for having gone fast. When that pilot comes home, his family is older, and he can never undo that, unless he sends *them* off on a fast trip and makes everybody older. It's kind of an Odyssey.

Earlier, I said data supports this strangeness. What kind of data?

The old textbook example is that of mu mesons, strange cosmic particles that are created 20–40 km above the Earth by cosmic rays hitting the atmosphere. These guys live only 2 μs, and should travel only 0.7 km at the speed of light before they die, but a lot of them reach the ground.

The best modern example is SLAC, the Stanford Linear Accelerator Center, where people make hard X rays by shining an 800 nm wavelength laser against an electron beam going at 99.8 % of the speed of light. But atomic clocks on satellites also show slightly different times than ones on Earth, just enough to validate Einstein, although part of *that* shift is due to our next topic, *General Relativity*.

General Relativity: The Final Weirdness

Einstein's other great discovery was that gravity bends light, which he called "general relativity." Did you know that? Lots of people I have talked to don't (Fig. 14). That looks sort of like a lens, doesn't it? It doesn't have to be a star—even Earth bends light. There's more to it than just bending light. Einstein showed that space itself distorts near a heavy thing and light is just taking the best way through when it bends, like Fermat's theorem [see Weird Reality]. The best practical application and proof of general relativity is gravitational lensing (Fig. 15). It's not a very good lens because of how it works—light rays bend more the closer they are to the star, while in a real lens light bends more at the edges, what you need to make a clear

Fig. 14 Gravity bending light (C. Phipps)

Fig. 15 Gravitational lensing (by NASA public domain)

image. In Fig. 15, gravity of that yellow spot in the center focuses light from a distant galaxy into that blue ring.

Just last week I heard a news announcement: some gravity lenses make several images and the light that you see may have also taken several different times to get here. So … scientists are able to see a supernova at several stages of its life, beginning, middle and end!

Another better-known example: black holes! When stars burn out and die, they sometimes leave behind a cinder with gravity so intense that light can't escape it. We talked about that a bit in *Weird Reality*. How does this affect you? Not much, unless one suddenly appears in your bedroom! That isn't going to happen, we think, but several years ago someone with a good imagination imagined black holes coming from laser fusion ….

Conclusion

Kepler and Newton changed the whole way we think about the Universe. They gave us the first true perspective on where we are located in it. In our solar system, planets move in elliptical orbits with one focus inside the Sun because it is so massive. If you're on Earth looking at Mercury for example, sometimes "Mercury is retrograde," because of the complicated motions in Fig. 3. It's like being on one Tilt-a-Whirl watching someone on another. You don't need Ptolemy's "epicycles," which he created to explain how we could be the center of everything and still see what we see. Although it might be comforting, we are *not* the center of anything. Not the solar system, not the Galaxy, not the Universe. There may even be other Universes that we haven't seen yet, because the light just hasn't got here yet.

Just by imagining that the Moon is subject to the same gravity that holds us to the planet, Newton was able to guess that gravity falls off like $1/r^2$ as you move away from a planet or any object, in order to fit what we know about its distance and speed. Three hundred years ago, he also understood there might be a problem with a force that is supposed to act instantly everywhere. Even today, we think gravity moves at the speed of light but there is yet no way to prove or disprove that.

Albert Einstein changed the way we think about reality itself. As something moves fast, electric fields and magnetic fields turn into each other, objects shrink, and mass zooms up. Worse, in some ways, lifetimes change. And, if you happen to pass near a heavy star, space itself bends.

In the next several chapters, we'll talk about practical results of Modern Science, from jet planes to rewriting DNA, from lasers to *real* transmutation of the elements!

Why Is the Day Sky Blue? Why Is the Night Sky Black?

On another planet, could the sky be green, red, or purple? After all, sometimes it's green in spots and stripes on a winter night in the far north around Norway. How would you feel about being assigned to a planet with a totally green daylight sky, if you were a space explorer and were stationed on that planet for a 2-year tour of duty? Why is our blue sky (Fig. 1) black at night (Fig. 2)? (Well, maybe gray if you live in Los Angeles, instead of Santa Fe!) Why is the sky at high altitude black at midday like it was for Felix Baumgartner before he jumped from a height of nearly 40 km? [More recently, Google executive Alan Eustace jumped from 41 km and made his own sonic boom!]

There are three parts to this question. If you let your mind wander and imagine different solar systems, the first part is: what color do you see on a piece of white paper at noon on a clear day? In our solar system, the answer is "white." In other solar systems, it might be red or orange. Some suns are red giants. That means no blue in your sunlight. If blue colors just weren't coming down from your sun, it would be pretty hard for your sky to be blue. It might be kind of yellow, and that would be it, so to speak.

The second part is, if you had the eyes of a bird and could see way down into the ultraviolet beyond blue on Earth, your brain might register a green sensation when you look up at the sky, because blue would now be in the middle of the range of colors you see. Of course, we'll never really know what honeybees see, unless …. Ah no, that's a different book! On the other hand, if the colors you could see included infrared, our blue sky might seem orange to you. Anyway, my point is that what you see depends on your eyeballs, and your brain, as well as what's there.

The third part is, why do you see sky color at all when sunlight comes through clear air? Why is what you see when you look up at the sky on Earth different from what Felix Baumgartner saw when he looked up?

What is clear air made of? Well, ignoring smog, dust and trace gases like carbon dioxide, it's mostly nitrogen and oxygen. In fact, it's mostly nitrogen. Nitrogen is made of molecules, pairs of nitrogen atoms. Light bounces off those molecules, just a little bit. We're lucky because, if it bounced off a lot, it would be dark all the time

Fig. 1 Blue Sky above Leadville, CO (C. Phipps)

Fig. 2 Black Sky from orbit above South China Sea (NASA public domain)

on the ground. Most of the light would bounce off high in the atmosphere before it got to us.

That's exactly what happens on Venus, where the air is 97 % carbon dioxide and there's so much of it that it acts like a white blanket. Air pressure on the surface is 92 atmospheres, so it's a very thick blanket. Most of the sunlight bounces off so the planet looks like a jewel to us here on Earth.

Fig. 3 Yellow sky on Venus, imaged by the Russian Venera spacecraft (http://i43.tinypic.com/iepdzk.jpg)

But for the Venusians on the ground, it's dark and miserable: dark orange sulfuric acid clouds above a sea of liquefied carbon dioxide, and hotter than hell: 470 °C, hot enough to melt lead. Air pressure on the surface is 92 atmospheres, so it's a very thick blanket. The Russian Venera craft survived this for 23 min, believe it or not (Fig. 3). You've heard of the greenhouse effect. Carbon dioxide, methane and some other gases trap heat—that's what we're worried about on Earth as more and more cars, powerplants, and factories burn coal, oil, or gas and put it into the air. On Venus, light does reach the surface, but it all stays there. Heatwise, like what they say about Las Vegas, anything that happens on Venus, stays on Venus.

Back to the nitrogen molecule: light bounces off it a little and the blue part of sunlight bounces off a lot more than the red part. If you're a blue ray the color of a blue LED, the chances you bounce off a nitrogen molecule are 16 times what they are for a ray the color of a red LED.

Why is that, and why are the chances of bouncing so small? You know that light is just a very short radio wave, right? And the molecule acts just like a tiny antenna. Because of its tiny size, it's not very good at picking up and rebroadcasting any of the light we can see. Like I said, that's a good thing. The shorter the wave, the closer the molecule is to the right size, and the better it is at picking up and rebroadcasting that wave. The nitrogen molecule (Fig. 4) is 1/4000 the size of a blue photon's wave but 1/8000 the size of a red one and that makes a big difference: it rebroadcasts blue better, although neither color very well.

Sunlight contains a lot of blue. We can see blue. Nitrogen rebroadcasts blue better, and that is why the sky is blue. That's all there is to it.

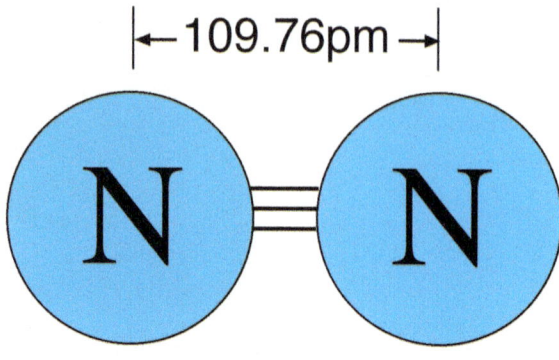

Fig. 4 A nitrogen molecule (C. Phipps)

J.C. Maxwell

Woops. I heard someone say they didn't know light is a short radio wave. So let's go back to that. It's only been about three times the age of your grandpa since people knew that at all. In 1862, Jim Maxwell figured out that radio and light are just the same kind of waves with different length, going at the speed of light. For more, see the chapter on *Electromagnetic Waves*.

Now, back to planets with green skies. Now that we understand sky coloring, what would it take? First, a sun that didn't have much blue. Our sun is like a very hot electric stove element. If you could turn up the power on the front burner of your electric range so that several million watts were going through it—what would happen? Well, you know the answer: it would turn red, then orange, then yellow, and blow up in a shower of sparks. If you were standing far enough away, it would be fun to watch. But if it was made of some magic element with a high melting point and kept away from oxygen, what would happen? Have you ever looked into an arc light? Don't do it, it will damage your eyes. But if you could look in there when the guy starts one up, you would see two carbon rods, sharpened like a pencil, connected to a lot of power, brought together to make a big spark and then pulled apart a bit so that a plasma arc forms between them in the gas. That arc gets to thousands of degrees, and goes on past yellow to make white light, which is why they use it in searchlights.

What is white light? It's light that has all the colors we can see from red to blue in it. Get a prism from the high school science lab, or buy one from Edmund Scientific, take it out in the sun and look at the rainbow of colors it makes. There they all are, red on one end, blue on the other and orange, yellow and green in the middle. You could (and Isaac Newton did) take a second prism and put these colors back together into white light [*Optics*].

There's a peak intensity in the spectrum of a hot object (Fig. 5), and the peak is bluer and bluer as the object gets hotter. Max Planck, the physicist who originated quantum theory, and others figured out the law that governs that spectrum (Fig. 6).

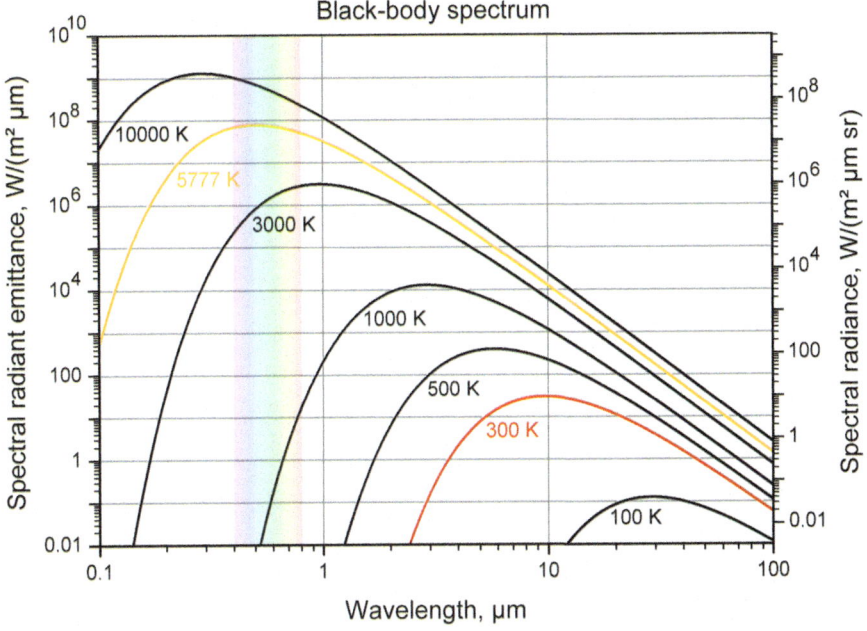

Fig. 5 Spectrum of a hot body vs. wavelength at several temperatures. The peak moves to the *red* as temperature goes down (Sch, GNU free documentation license via Wikimedia Commons)

Fig. 6 Planck (public domain)

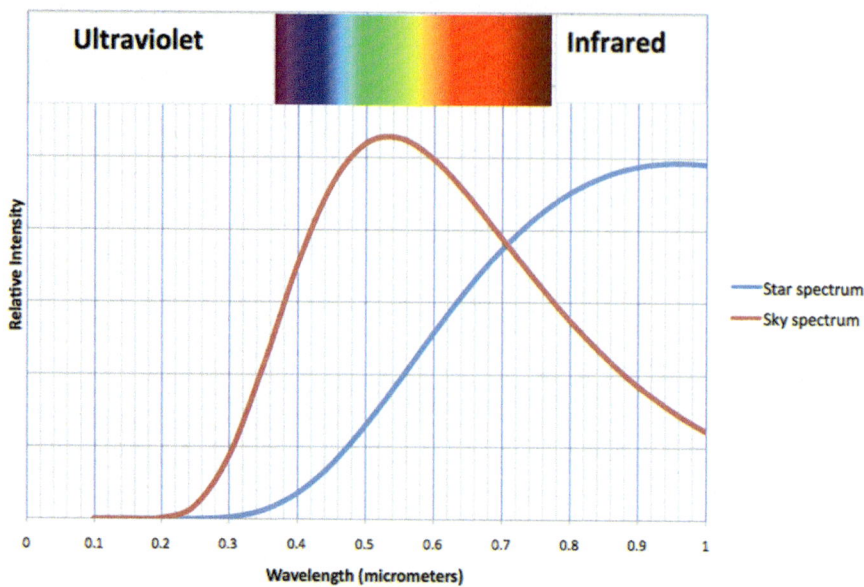

Fig. 7 *Blue curve*: spectrum of an orange sun with a 3000°K surface. Its peak is out there in the infrared where you wouldn't see it. *Red curve*: Spectrum of the sky. It peaks in the *green*! (C. Phipps)

He doesn't look too happy, does he? As a matter of fact, he got the Nobel Prize in 1918 for discovering the quantum even though he didn't claim to understand it. His famous formula for the spectrum shown in Figs. 5 and 7 [called blackbody radiation] was "an act of desperation," he said. To him, the idea of quantization was only "a purely formal assumption … actually I did not think much about it." He was in good company—even Einstein didn't believe in many aspects of *Quantum Weirdness* up to the day of his death. Maybe that's why Planck looks sad.

If we go to a planet in a different solar system where the sun is cool, on its way to being a red giant, the blues will disappear first, leaving greens and yellows on one end of the rainbow. Let's say the planet has an atmosphere like ours. The planet had also better be a lot farther from its sun than we are, or it will already have been gobbled up, because red giants are not only red, but giant. On our new planet, the air will still retransmit the shorter wavelength colors better than it does the red ones, but now it only has green and yellow to work with on the short end, so that will be the color of the sky. A sun that has cooled to a temperature of 3000°K (instead of 5500°K like our sun) ought to be just perfect for green skies. And now, if I ask what the sky spectrum looks like after scattering blue 16 times better than red, Fig. 7 (blue curve) shows what we get. A green sky! Did you ever see the movie *Forbidden Planet*? Altair 4 had green skies.

What are degrees Kelvin? Just degrees above absolute zero, a more useful thermometer for physics than the one we use. See *Metric System* to understand these units.

At the beginning, I mentioned red skies. How would you get that color? When the sun has cooled to 2300°K, the sky would be cherry red. For violet, to tell the truth, we have that already, especially under ozone holes, and that's what tans or burns your skin. Our sky is really violet, but you can't see it. Only a honeybee can.

Dark Night Skies

Does it seem strange to you that the night sky is dark (Fig. 8)? (Or, sort of gray and murky if you live in New York or LA) What I mean by that is this: if you look out into the universe, it's mostly dark with a few spots of light. Did you ever wonder if that makes sense? The Universe is almost infinite, right? If you can see infinitely far and there are an infinite number of stars the sky ought to glow like daylight, whether it's day or night.

Or not? This was a mystery that bothered people for a couple of centuries. It's called Olbers' paradox, although he got the wrong answer. Actually, Bill Thompson, otherwise known as Lord Kelvin, figured it out. And so did Edgar Allen Poe even earlier in an elegant paragraph [1848]:

> Were the succession of stars endless, then the background of the sky would present us a uniform luminosity, like that displayed by the Galaxy—since there could be absolutely no point, in all that background, at which would not exist a star. The only mode, therefore, in which, under such a state of affairs, we could comprehend the voids which our telescopes find in innumerable directions, would be by supposing the distance of the invisible background so immense that no ray from it has yet been able to reach us at all.

Fig. 8 Night sky (ESO/H. Dahle (http://www.eso.org/public/images/potw1333a/) Creative Commons License, via Wikimedia Commons)

In brief: the Universe is NOT infinite! It's only 14 billion years old, so we can only see stars within a distance of 14 billion light years [130 billion trillion km!] and, in the space we can see, there are not enough stars to fill the whole sky with light. If there is light from farther away (but there won't be, unless it's from another Universe!), it hasn't had time to reach us yet!

Not only that, stars have only been lighting up for the last 8 billion years, not 14. Not only that, the ones that are there burn out in a limited lifetime, about four billion.

Not only that, the Universe is expanding. You may have heard that. What it means is that the further a star is from us, the faster it is moving away. Its light gets shifted toward the low frequencies (red and beyond) just like a train whistle when the train is moving away. A lot of stars and galaxies are going away from us so fast that the "redshift" of their light moves it down into ranges we can't see! How much of what we do see was shifted down from the blue and ultraviolet to the visible when we look at distant galaxies? To answer that we'd need to guess our galaxy is typical and then measure the amount of UV emission in a sky survey. But because sky surveys are normally done on Earth and we can't see the UV through the atmosphere, we don't really know. From what we do know, there may not be so many of those UV stars.

So: If I ask at what distance the redshift is so great that what began as blue has become red, and red invisible (a redshift of 1), it's about 1.9 billion light years. Think of the universe as a sphere, and the radius of the "lit Universe" about 8 billion light years. Then, you might think that if it's uniformly populated, we can only see light from less than 2 % of it [$(1.9/8)^3$] or, 98 % of the universe that was born visible is redshifted so we can't see the light! Of course, this analysis is too simple, and we're not really at the center of everything, but here is another reason the night sky is dark.

Finally, only 5 % of the mass of the Universe is anyway normal stuff you can see, like you and me and the stars (baryonic matter). The other 95 % is dark matter, radiation, vacuum ("dark") energy and whatever else. If this weren't so, there would be a lot more stars for us to see [*Weird Reality*].

So there are a bunch of reasons why the sky is not blazing at night!

Aurorae

To finish this off, what about the green or blue flames in the night skies you see sometimes above Earth's polar regions (Fig. 9)? Aurorae are a whole different thing. The sun not only sends out light, but high speed electrical particles that are guided into the air above the poles by the Earth's magnetic field lines. These bend in toward the poles, and you can see the results when it's dark if you're up or down there. While still high above the poles, the electrons plow into the air atoms they find and tear them apart. When these get their electrons back they glow with certain colors depending on the gas.

Fig. 9 Aurora Borealis (Jerry Magnum Porsbjer, GNU free documentation license via Wikimedia Commons)

Conclusion

Yes, you could get a green sky, or a red sky, on another planet. We already have a violet sky. It depends on what colors your sun puts out, what gases are in the atmosphere, what our eyes can see and what our brains do with that information. The Universe is expanding, but it is finite, so some light hasn't reached us yet. Distant stars redshift so you can't see them. Not very much of the Universe is even *stuff*. These things taken together are the reason for dark skies at night. Auorae are another way we sometimes get a green sky on Earth.

In the next chapter, we'll talk about machines.

Machines

People say Man is a toolmaker. I say Man is a maker of *machines*. Anybody can make an axe or a hammer, and these were very important inventions at the time. But, a *steam locomotive*? Or a *windlass*, to amplify what he can lift tenfold, or a *car*, to let him travel ten times as fast as he can run, or a *sailing ship*, to let him travel clear across an ocean he could never swim—that's something! Along the way, there were some many ingenious lesser inventions to make those things a success. And there is already one machine that can duplicate itself! Do you know what it is? I'll tell you at the end of the chapter. As usual, we can't talk about them all, so we'll hit the high spots.

Force Multipliers

Those humongous stones that were lifted and put in place to make the Pont du Gard (see *Rome*) were put there with a windlass (Fig. 1), or with a block and tackle (Fig. 2) or a combination of these. Do you understand how a windlass works? Of the two, the windlass is the most ingenious invention. The Chinese had these. As the rope on the right coils up, the rope on the left uncoils, but a little less with each turn of the crank. Each turn raises the load a tiny amount proportional to the *difference* of diameter in the two sections. Force times distance is a constant, so if you raise the load a little bit while you turn the crank a lot, you can lift a very heavy thing.

A block and tackle is the most practical for multiplying human lifting power a lot, and you don't tangle the ropes. Figure 2 illustrates how you lift four times your strength with a double pulley. Each pulley pair double the force. You can imagine lifting half a ton all by yourself with five pulleys on each end of that block and tackle. Then, you need ten times the length of rope as the distance you want to lift. If you ever stopped to look at a crane, or past the glass in a transparent elevator, you know that people use many pulleys at top and bottom to let an ordinary motor lift tons.

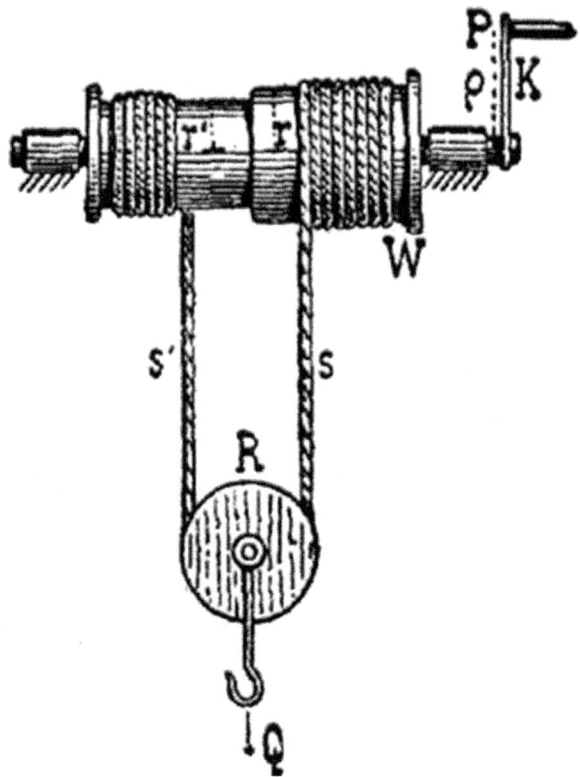

Fig. 1 Windlass (Otto Lueger 1904 public domain via Wikimedia Commons)

You can also drive one of these with a waterwheel (Fig. 3). I'm including that picture to illustrate two ingenious inventions at once. The Romans had waterwheels, as did the Chinese and later, Islamic engineers.

Up until very modern times, water power has been used for everything from grinding flour to crushing rocks. It all depends on whether you have running water, or not. If you don't, then a poor donkey or ox has to spend his life running around in circles.

External Combustion Engines

Think "steam engines." You build a fire under a boiler (so, external combustion), make steam, and use that to do something.

The first one was invented by Hero (Heron) of Alexandria in the first century AD (Fig. 4). You can see how it worked by looking at the picture. This was the first steam turbine! He never used it for anything but amusing people.

The first successful steam engine is shown in this picture of James Watt's engine (Fig. 5). He had access to the best iron workers in Scotland, who knew how to bore

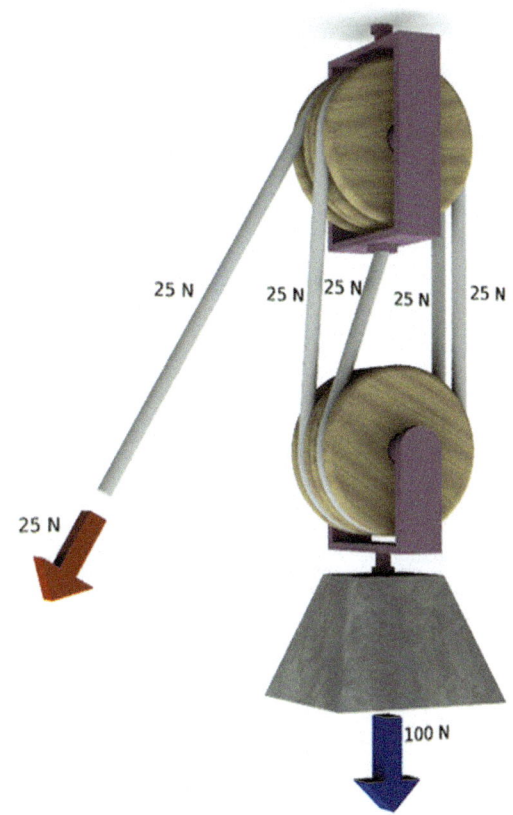

Fig. 2 Block and tackle (Otto Lueger CR, GNU free documentation license via Wikimedia Commons)

cannons accurately, and that was a key to making pistons and cylinders that fit each other. This was a key to operating efficiency. If you're having trouble understanding Fig. 5, Watt was injecting steam into the cylinder and then condensing it by spraying cold water in there, so a vacuum pulled the piston down, believe it or not. Well anyway it was the eighteenth century.

As I said in *Metric System*, Watt undersold and overproduced when he established the horsepower unit, making sure his engine could do the work of two horses. Always a good plan for an entrepreneur.

Today of course, we squirt very hot, high pressure steam into that cylinder, and let it cool *as it does work* on the piston.

There are two types of modern steam piston engines—"water in the tube" and "fire in the tube."

To me, water in the tube makes more sense because it's easier and less dangerous to contain pressure in a tube. But the old railroad locomotives still running around the Rockies use the other approach, which is why they have all those bolts around the flanges on the ends (Fig. 6). If you want to ride behind one, check out the Durango-Silverton or the Chama-Antonito lines.

Fig. 3 Water wheel mining hoist (Public domain, *De Re Metallica*, 1556)

Fig. 4 Heron's Aeolipile (Public domain via Wikimedia Commons)

Industrial Revolution

It's hard to overemphasize the importance of the steam piston engine. The first engines were simply used to pump water. James Watt realized you could turn a wheel.

Voila! There you go down the railroad track, selling and shipping goods all over the world as fast as you can feed coal into this thing. There goes your generator, mill or steamboat. There go a whole fleet of steamboats and steamships, with cannons so you can control the world. These machines make other machines, a technological *Exponential*, such as we have seen often in this book. There was even a steam automobile, the Stanley Steamer.

Carnot and Power Generation

In 1824, the French scientist Nicola Carnot showed that thermal efficiency of an engine goes like $\Delta T/T_{hot}$ where ΔT is the difference between input and output gas temperature.

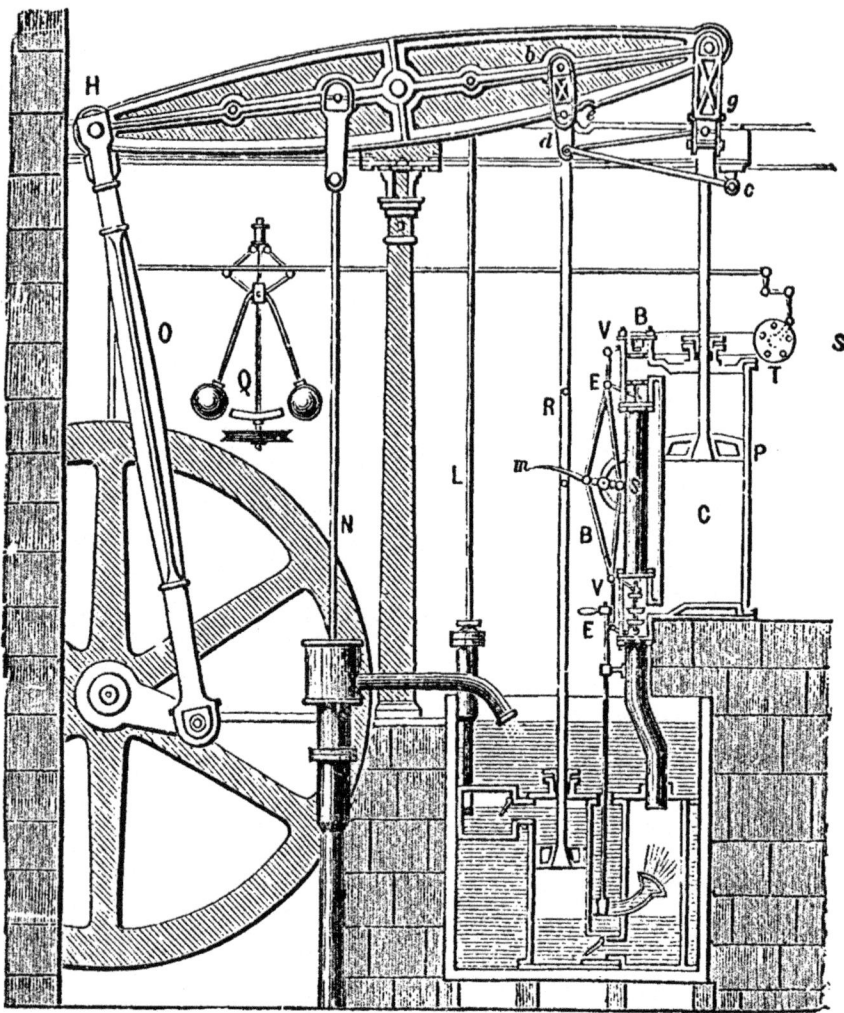

Fig. 5 The Watt engine (1784) (R.A. Sallinen III Creative Commons license via Wikimedia Commons)

Even 50 % is hard to get. A power station might get 40 % and there's another 20 % loss in transmitting power down wires through transformers to your home. Put that all together and you've got 30 % from fuel to electricity at your toaster. That's a good number to keep in mind when you think about solar power on your roof being 15 % already, and you don't put 70 % of every joule you use into the planet's air as wasted fuel.

Fig. 6 A 1940s Union Pacific "Big Boy" Locomotive (American Locomotive Company, about 1940) (RFM57, Creative Commons license via Wikimedia Commons)

Improvements

People realized right away that a single-cylinder high pressure steam engine wasted more precious water and coal than necessary. What if you ran the exhaust steam from the first cylinder into a second cylinder and let it push on a second, larger piston? Look closely at the front of the engine in Fig. 6, just behind the "cowcatcher", and you'll see two cylinders. This is called a compound engine. Some engines used three stages. A second improvement came by realizing you can continue to heat that steam until it's a dry, hot gas before using it. This is called superheat. In locomotives, 180 °C was a typical figure, but as much as 400 °C was used. Steam locomotives got up to about 21 % thermal efficiency.

The Fig. 6 engine really was a big boy. It weighed 762,000 lbs. [346 t], pulled 135,000 lbs. [61 t], developed 7160 horsepower, and when moving at 70 mph [113 km/h] on its 68-in. [172 cm] wheels, used 9.6 t/h of fuel.

Miles per gallon weren't great: those numbers tell you it had a fuel consumption of 124 kg/mile. If it used fuel oil, which it didn't do very well, you would say it got 0.03 mpg! But of course it could pull 6500 t down the track! If you compare that to a semi truck hauling 64 t at 1.7 mpg, 100 trucks going down the highway at a total 0.017 mpg are collectively two times *less* fuel-efficient than the Big Boy, doing the same job.

It operated at 300 psi. That's 211 t force/m^2 to put it in slightly non-metric units that emphasize the terrific force those bolts are holding down! The downside for steam was the 10,000 gallons per hour of water the Big Boy blew into the atmosphere. In 1940, each engine cost $225,000.

Today we use diesel-electric engines on trains, which can get 36 % thermal efficiency and do not require much water. But wait: recently, South African engineers claimed that well-designed steam is cheaper to maintain than diesel! There may be a future for steam.

Stirling Engine

This one is a bit harder to get your head around, but it's neat (Fig. 7). The new thing is the displacer piston, which acts as a barrier between the hot and cold gas in the cylinder and in some versions may be free, not connected to anything. Like everything, its efficiency depends on how hot the hot end is, but it can be 30 %. You can heat it with anything, including the sun. It has no exhaust or intake valves, and never has to endure explosions and erosion inside.

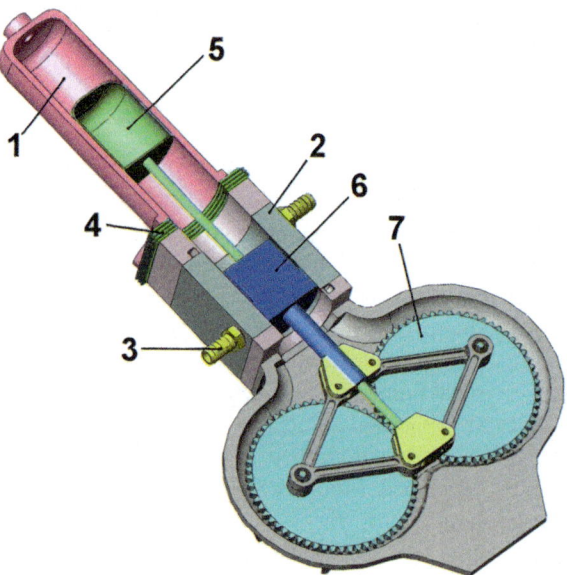

Fig. 7 Stirling engine. You apply heat at (1), bring coolant in at (3).m (5) is the displacer piston and (6) is the power piston (BetaStirlingTG4web.jpg, GNU free documentation license via Wikimedia Commons)

Steam Turbines

Steam turbines are another matter. Still external combustion, but huge, very efficient, stationary engines. Charles Parsons made the first one in 1884, to run a 7 kW electric generator. Today, they're built up to 1.5 GW (Fig. 8). You can think of it as a huge waterwheel or fan, with different fins. These use steam at 500 °C temperature and 8.3 MPa [1200 psi] pressure, so 50 % thermal efficiency is easily achieved. This is very likely the front end of the generator that brings power to your home.

Using the same idea as in the compound piston steam engine, these use a high and low pressure turbine in tandem.

That reminds me of an embarrassing story: quite a few years ago, as an Engineering Duty midshipman in the Navy, I was stuffed into the engine room of a troop transport for a training cruise. When the Captain rang the bell, I was supposed to open the throttle to one of those compound turbines. The Chief assumed I knew all about it, and I didn't let on. He reminded me I needed to get 60 rpm or whatever out of the propellers when the signal came, and then disappeared for coffee. The bell rang. In front of me was a giant chrome wheel with a knob. I turned it counterclockwise as fast as I could. The pressure dropped, the propeller speed picked up and got stuck 30 rpm. The Captain was irate because we were in port moving away from the dock, he needed to get what he asked for, fast, and he let me know. At the critical moment, Chief returned, somewhat wild-eyed and opened another valve. "You gotta use the high pressure stage, sir!" That event engraved that fact in my memory.

Fig. 8 Steam turbine (Siemens press photo, Creative Commons license via Wikimedia Commons)

Internal Combustion Engine Zoo

Otto Cycle

Did you know that's what runs your car? You can probably figure out the parts (Fig. 9). The only difference with a steam engine is that the fuel and air mixture explodes. S is the spark plug that ignites the mixture, which came in through the intake valve I, and E is the exhaust valve that lets out the burned gases. The explosion drives the piston P down, pushing the rod R to turn the crank C. There are several pistons connected to the crank, one after the other in a line. Its energy efficiency is 38 % at most. Power to weight ratio is about 0.36 horsepower/kg.

Wankel Cycle

That sounds a bit threatening, but Dr. Wankel had a dream here. I was the proud owner of a Wankel Mazda for 2 years back in the 1970s (Fig. 10). That triangular middle part moves in an eccentric path around the central shaft and, via gears, turns it. There is no crank. You have to see a model or animation to understand. It topped out at 125 mph without any effort and the loudest sound at all speeds was the outside air whizzing by the cab. It felt more like a steam turbine than a gasoline motor. It was great when gas was 25¢/gallon. Its problem is that the average shape of the "cylinder" is a banana rather than a cylinder, giving a lot more surface to volume ratio, and more heat loss, therefore lower mileage. I think I got 15 mpg.

You can't imagine how sad I was when I wrecked it one night, distraught over a beautiful young Armenian lady.

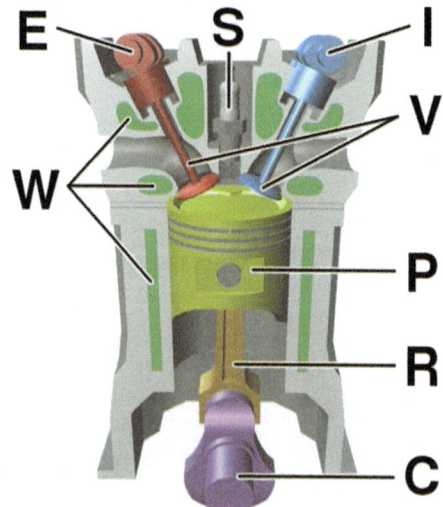

Fig. 9 Modern gasoline engine (Wapcaplet GNU free documentation license via Wikimedia Commons)

Internal Combustion Engine Zoo

Fig. 10 Wankel rotary engine (Softeis, German Wikipedia. GNU free documentation license via Wikimedia Commons)

A Wankel engine is very light, and gets about 2 horsepower per kg. Its only problem was *sealing* all those surfaces at the tips and sides of the rotor to keep hot gases out of the oil and vice versa.

Radial Engine

For aircraft, you could never carry around the massive weight of a standard gasoline engine with its long crankshaft. So, here is a genius idea that achieved 2.6 horsepower/kg (Fig. 11). It's hard to find a good illustration of the entire engine outside its case. You have to imagine pistons and cylinders at the ends of each connecting rod. The point is: there is only one crank. But note: the radial engine has the same advantage as the Wankel! Just one off-center piece turning the shaft!

Diesel

Originally, Dr. Rudolf Diesel wanted to build an engine that would use a mixture of oil and coal dust. But his main idea was to obtain much higher efficiency by burning fuel at higher temperature and pressure. To do that, he increased the compression

Fig. 11 Radial engine connecting rods and crank (TSRL, Creative Commons license via Wikimedia Commons)

ratio to such a high value that the fuel ignites itself. Diesels routinely compress the air and fuel mixture as much as 22 times, while gasoline engines normally use half that. Because they're hotter, diesel engines get 55 miles per gallon even without a hybrid design. To do that, the engine casing has to be very strong and very heavy, but diesels do get about 0.9 horsepower/kg because of their efficiency.

Hybrid

Hybrids are a type of car, not a type of engine. The gasoline/electric or diesel/electric hybrid car is all you need to save about half of the fuel we burn to travel around on the ground. My Prius gets 45 mpg all the time using a very ordinary little gas engine. The main savings is from turning off the engine at red lights, and using an electric motor/generator to store the energy normally wasted in braking. One of the cleverest things about the design is that there's no clutch and almost no brakes. I have gone 125,000 miles on high voltage battery and one set of brake pads.

Tesla

Now, what about all-electric cars? Obviously you need lots of good batteries that are cheap. Considering that the energy density of gasoline is 42 MJ/kg and that of lithium batteries about 42 times less [see Electricity], it is to Elon Musk's credit that he's made such a success out of the Tesla car (Fig. 12). Electric cars have been around since the 1880s, but the Tesla is a beautiful, high-performance thing and EPA says it costs 4.5¢/mile to drive.

Internal Combustion Engine Zoo

Fig. 12 Tesla Roadster (Tesla Motors, inc. copyrighted free use license via Wikimedia Commons)

Fig. 13 The 550 horsepower STP-Paxton Turbocar (The 359 GNU Free Documentation License via Wikimedia Commons)

Turbine Engines for Cars

You might wonder why turbines haven't been used in cars, considering that they can achieve 10 horsepower/kg! At the Indianapolis races in 1967, Andy Granatelli tried that. He ran an STP-Paxton turbocar, which was so good it was banned (Fig. 13). Just look at that thing! Parnelli Jones came within a few miles of winning the Indy 500 by an unheard-of margin. He was a whole lap ahead of the second car when the transmission broke. There are people who are suspicious about that result. I know I was, at the time! In any case, the rules were changed so that turbine cars couldn't compete after that.

The Differential

Now here's something you may not think is ingenious, but I do! (Fig. 14) Why? It manages to turn your drive wheels [on the shafts (1)] while at the same time allowing them to rotate independently. If you didn't have that, the axles would twist off

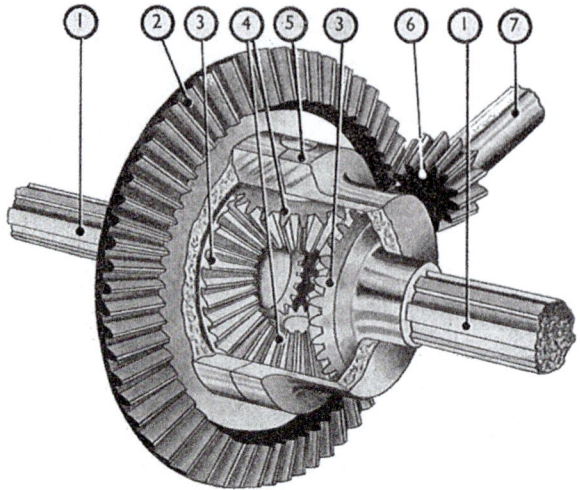

Fig. 14 Differential (Andy Dingley public domain via Wikimedia Commons)

Fig. 15 South-pointing Chariot (Andy Dingley Creative Commons license via Wikimedia Commons)

or rubber would slide as you go around a corner! Did you ever think about that? Again, it's hard to find a good drawing, but the driveshaft (7) turns that big gear (2) which is connected to the housing that holds that smaller pair of gears (4) perpendicular to the axles, which are driven by the gears (3). If one wants to turn faster than the other, that's fine! As an invention, it goes way back. As the Chinese cart in Fig. 15 goes around corners, the driver always points in the same direction!

Windmills and Renewable Energy

I don't want to snow on your solar panels, but guess what: windpower is a lower cost source of renewable energy, installed, than solar electric, *by a factor of four*. "It can't be!" you say. But it is 1.4 cents per kwh vs. 5.6, and the reason people don't use them more is NIMBY (Not In My Backyard). I think they're beautiful [see Fig. 7 of Jet Planes], but a lot of folks don't. You'll notice the mill in that figure is out in the ocean where it can't bother people. Here in New Mexico where it blows a lot, a neighborhood could power itself from *one* of them, which costs about 3 M$ installed and makes 2000 kW. Sounds like a lot, but it's $1.50 a watt. Compare that to $4.90/W for installed solar! New coal plants are about $2.10/W. I see windmills all over southern France, as I write this.

Yes, you need batteries to get through the night if you want to be completely off the grid, and that adds to cost. Or, you can just use a combination of the power company and renewable, which makes them crazy but suits you just fine.

Another use for windpower is desalination, and this is important. Here, you do not need to store electricity, only fresh water. I am working on that one!

Inverse Machines, the Servel and Liquid Helium

If you can make an engine that turns energy from hot and cold reservoirs into work, why not go the other way and turn work into a refrigerator? I hope you do know that has been successful! The electric motor in your fridge compresses gas which goes into a radiator in the back (touch it—it's hot, right?) where it cools off and then is allowed to expand again and cool the inside. That part is the very same thing as when you use up a can of butane on your camper stove and it ices up. When my dad was a kid, they just delivered blocks of ice to your door. Those came from a plant downtown which, back then, used ammonia for a refrigerant. Now for something really surprising.

In those same old days, people had Servel refrigerators. If you have an RV, you probably still have one. It uses a flame to refrigerate stuff! This is neat if you don't have electricity. How the heck do they do that?

A flame heats a mixture of ammonia, water and hydrogen. Yes, hydrogen. Ammonia bubbles pump slugs of ammonia liquid up through a tube, like those Christmas bubble lights, so it can then drip down through coils inside your fridge. Before it drips, it has to get rid of the heat, which it does in the cooling fins outside your fridge. The hydrogen is there to help the ammonia evaporate quickly in the tubes that are inside your fridge. That does the cooling. All driven by heat and gravity without a single mechanical moving part. Who thought this up? That wonderful inventor, Mike Faraday that we mention in *Electricity*, in the early 1800s!

Speaking of unusual fridges, how *do* you make *liquid helium*? It's not easy, because LHe is just 4°K above absolute zero, which is *minus* 460 Fahrenheit. Heike Kamerlingh Onnes (HKO) [*Weird Reality*] needed it, so he figured out how to do it in 1908 (Fig. 16). To start, you put a whole bunch of refrigerators in series, each

Fig. 16 HKO and his liquid helium factory (Dirk van Delft, Museum Boerhaave, Leiden)

with different refrigerant liquids, ending with liquid hydrogen. Then you let the hydrogen boil to cool the helium. *Or,* you let it do work on a little piston engine— I've seen that.

Electric and Laser Refrigerators

Not only that ... you can refrigerate things with an electric current! This is called a thermo-electric cooler. Because it's fed with DC, it doesn't even hum (Fig. 17).

Umm ... not only that—have you heard of the laser refrigerator? I would bore you if I explained how it works, but it does. Los Alamos called it the LASSOR for Los Alamos Solid State Optical Refrigerator.

The Self-Replicating Machine

My high school shop teacher in pointed out that the lathe is the only machine that can make itself. Now, come on, you say—what about 3D printers? Well, they haven't made one horsepower electric motors yet that way, although the day may not be far off. But back then, this was a fascinating statement. Figure 18 shows one from Mr. Kester's time.

You *can* make almost anything with it.

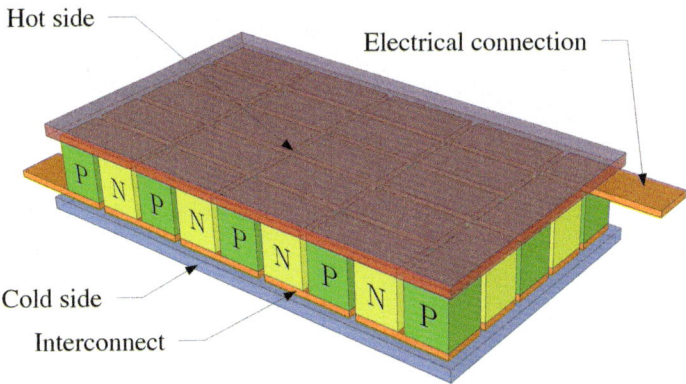

Fig. 17 An all-electric cooler. P and N refer to different kinds of semiconductor. Plus and minus determine which side gets cold and which hot (Michbich, Creative Commons license via Wikimedia Commons)

Fig. 18 Lathe (Ukexpat Creative Commons license via Wikimedia Commons)

Other Machines

I can't cover them all. *Jet Planes*, *Lasers*, *Drones*, Bombs (*Nucleonics*) and *Rockets and Satellites* deserve their own chapters, and they have them.

Conclusions

This chapter has been a zoo full of machines people have invented to help them do things they couldn't do by themselves. Not everyone thinks the results are great for our society, but not everyone agrees on anything. Cars clog freeways and emit pollution, but they can be improved and they get you places fast. Some have really exciting futures! See what you can do to change the world for the better!

Lighter Than Air Flight

Lighter than Air flight is just filling a bag of some sort with gas that is lighter than air. Then, the difference lifts it, like Styrofoam in water, and you too if you're attached and it is strong enough. Remember Fig. 12 of *Weird Reality*? If something is lighter than the fluid around it, it pulls away from gravity. This works for boats and for balloons. There are a couple of easy ways to fill a bag with lighter-than-air gas. The first is to heat it, because hot air is lighter than cold air.

Hot Air Balloons

Have you ever taken a ride in the basket of a hot air balloon? A few years ago I did, and it's fun! You can't believe how quiet it is when you're just 1000 feet or so above the noise of Albuquerque! I can only imagine how excited de Rozier and d'Arlandes were in 1783 when they did that the very first time in Paris (Fig. 1). This was the first successful human flight (where they didn't crash and break bones). The Chinese had hot air balloons (Kongming lanterns) in 200AD, but they didn't carry anybody. The time was 1783, and the balloon was made of silk fabric lined with paper, filled with hot air from an open fire (Fig. 1). The text on the official poster for the flight said "The aerostatic globe, the first to lift men through the air." Unfortunately for my praise of the French for their *Metric System*, this was before the Revolution, and the dimensions are given in feet (*pieds*), not meters. Not only that, the feet were about 1.07 of our feet. The capacity was 60,000 of those cubic feet, or 73,000 of ours. The balloon had a lifting power of 1700 modern pounds, equal to the weight of the balloon itself and its passengers. It would be another 15 years (after the Revolution) before France adopted metric.

To achieve that, the air temperature inside would have had to be 220 °F hotter than outside, or at least 290 °F, much hotter than people operate modern balloons.

It was June 4, so really very warm for a hot air balloon flight. On this first flight, there were no passengers. It stayed up for ten minutes, rose 2 km and traveled 2 km. Then they tried animals.

Fig. 1 Launching the Montgolfier hot air balloon from the Bois de Boulogne, Paris, November 21, 1783. Beautiful picture, no?! (Public Domain via Wikimedia Commons)

The first human passenger (de Rozier) was sent up on October 15 and, when he made it, de Rozier and d'Arlandes on November 21. Ben Franklin was there to watch it take off and fly 9 km to the Paris suburb of Butte-aux-Calles. They were so careful because they were afraid people couldn't survive the height—even though many people had climbed higher mountains.

Less than a year later, they flew 52 km in 45 min at an altitude of 3 km, with Joseph Proust on board.

Today, hot air balloon pilots carry propane tanks and a giant burner that can keep the balloon up for some hours (Fig. 2). At that time, limited fuel for the little brazier they carried onboard, and worry about fire, limited their range to a few km.

Fig. 2 A modern balloon with basket (HRae at English Wikipedia Creative Commons License via Wikimedia Commons)

Gas-Filled Balloons

How do you get away from hot air? Use a gas that is naturally lighter than air. What gases are lighter than air? Air has a molecular weight of about 29, so any lighter gas is great, for example ammonia (17), methane (16), helium (4), and hydrogen (2), going from the worst to the best lifting gases. Already by 1785, despite one lethal explosion, The French had already experimented with a hydrogen-filled balloon. This solved the problem of flight duration, but not the problem of controlling where the balloon goes. Ultimately, *de Rozier* used a combination of bags filled with lighter gas and hot air, which lasts longer and is more flexible. This is a design sometimes used today.

The first dirigible was also French, designed and flown by Jules Giffard in 1852, and powered by a three horsepower steam engine. It could go 10 km/h, but needed to fly in circles to demonstrate its capability, rather than trying to fight even gentle winds.

Dirigibles

This word simply means "steerable" in French. A dirigible is not necessarily a complex thing like the Hindenburg, with a rigid internal skeleton. That's a *zeppelin*, like the Graf (Count) Zeppelin … and the Hindenburg. Blimps, like the Goodyear one—from the British term "Class B–Limp" are just cigar-shaped bags, not rigid airships.

The Hindenburg was the largest airship ever built, with a length of 245 m, more than three times the length of a Boeing 747 (Fig. 3). It had a diameter of 41 m and a gas capacity of 200,000 cubic meters. That gave it a lift of 230,000 kg. It had 72 sleeping berths, 40 flight officers and men and ten stewards and cooks, and cruised at 125 km/h with four 1200 horsepower Daimler-Benz diesel engines driving its propellers. That meant it could make a transatlantic crossing in 2 days, compared to 4 days for the fastest cruise ship over a similar distance (even today). A better comparison might be New York to Rio in 10 days for the famous ship Normandie, vs. 4 days from Rio to Frankfurt with the Hindenburg.

It had a gorgeous dining room (Fig. 4). Not surprisingly, a one-way ticket to Europe from New Jersey was about $6000 in 2015 dollars, about twice the first class air or deluxe ocean cruise price today.

We're lucky that we have a lot of helium underground in Texas, because it doesn't explode. It's not much heavier than hydrogen, because helium is a single atom with atomic weight 4, while hydrogen (atomic weight 1) has to exist as a molecule of two atoms with molecular weight 2.

Everyone remembers the fate of the Hindenburg. Germany had to use hydrogen because they didn't have much helium, and they couldn't buy it from us because we

Fig. 3 The Hindenburg (1937) (public domain)

High-Altitude Helium Balloons   163

Fig. 4 The Hindenburg dining room in 1936 (O. von Stetten, Deutsches Bundesarchiv, Creative Commons Germany License via Wikimedia Commons)

named it a strategic war material in 1937. Amazingly, 65 of the 100 people onboard this last flight survived this fiery crash. It made more than 60 flights before the disaster. There's never been anything like it.

Even today, people who think a transatlantic cruise takes too long but would like to travel to Europe in style over a couple of days hope for a revival of the Zeppelins.

Helium, in Texas?

Near Amarillo is a big underground deposit of helium mixed with methane. It's one of the few places on the Earth where you get lots of it. What is it doing there? Radioactive decay of things like uranium makes alpha particles, which are just charged helium nuclei. You can also find it underground in north Africa, Russia, Kansas, and the Oklahoma Panhandle, but Texas is the main place.

A personal warning: as funny as it is to breathe a lungful and talk very high, it's also a great way to kill yourself because your lungs are a diffusive membrane. A lungful of helium will cause the oxygen that's already in your blood to quickly diffuse out of it, just like a lungful of oxygen makes that diffuse in. *You can't detect that something is wrong until you pass out*, and that can be in just a few breaths.

High-Altitude Helium Balloons

Today, helium-filled balloons are a very important research tool for NASA (Fig. 5). They can reach 40 km (130,000 ft) altitude. You may have seen these hovering bright and silent in the twilight over your city, and wondered if it was a UFO!

Fig. 5 NASA helium-filled polyethylene research balloon at 40 km altitude (NASA public domain)

Distance Records

Right now, the distance record for non-powered helium balloons is held by the January 31, 2015 flight of "Two Eagles," 10,696 km from Saga, Japan to Baja California, in 160 h, 37 min. Do the math and you'll find that's 67 km/h. The pilots were Troy Bradley and Leonid Tiukhtyae.

Spying

Even during the American Civil War, one of the first applications people thought of for balloons, as for any new invention, was spying on "the enemy." Suddenly, you have a vantage point high above opposing army forces, better than any hilltop, and you can see what they're planning to do. Even today, we have used them in Iraq and Afghanistan. And, in a famous 1947 incident, a high-altitude balloon used by *Project Mogul* crashed near Roswell, starting a generation of "UFO" sightings. This was a project to listen for telltale sounds of Soviet bomb tests. As I said a couple paragraphs ago, these can be really eerie at twilight!

Conclusions

Lighter-than-air craft are the earliest flying machines, and they're still very useful. Not only that, but they were once the most elegant way to cross the ocean. Not as slow as a steamship, but still sweet for those of us who feel it's uncivilized to have the sun come up at 2 a.m. in Paris.

Electricity

It makes everything from iPhones to Teslas run, but what *is* it? What are AC and DC? Are some things AC/DC? How do electric motors work? What do the words volts, amps, ohms, and watts mean, what can they do? You really should know the difference to be safe and also not to worry unnecessarily. Can a 12-V battery electrocute you? We'll get into all these things in this chapter.

What is *electricity*? It's just the flow of electrons. Inside a good conductor like copper, there's a whole sea of free electrons floating around and, with a few volts, you can make them go clear through and out one end of the wire where they were born into a toaster or an electric motor and do some work. The ones that go out are almost instantly replaced by new ones coming in. The wire doesn't own them!

Volts, Ohms, Amps, and Watts

Your vacuum cleaner probably says "120 V" in the USA, and 220 elsewhere in the world. Also, "8 amps" or so in the USA, where amps are sometimes advertised like a measure of capability in vacuum cleaners these days. What are these things? Can you be electrocuted touching the 12 V battery in your car? (No) Can you burn yourself badly if you short it out with a wire? (Yes!) And what are watts? Or, for that matter, ohms? Do you care? Just in order to vote intelligently on nuclear power plants and stuff, you should care, and know.

All these things are named after early electrical scientists: Alessandro Volta, Georg Ohm, André-Marie Ampère, and James Watt.

Figure 1 is a little mnemonic device a teacher of mine once gave me. Table 1 explains it in words, but a picture, as we know, is worth a thousand of them. It says volts over amps is ohms, volts over ohms is amps, and amps times ohms gives volts. The table adds one more thing: power is volts times amps and is measured in watts. Power is also the rate at which we use up — and sometimes waste — energy, and the precious fuel that makes it.

Fig. 1 Volts, ohms, and amps (C. Phipps)

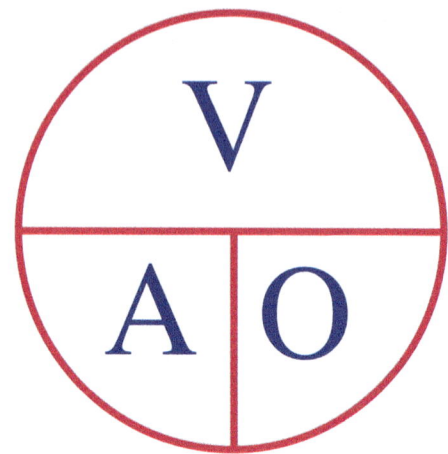

Table 1 Volts, ohms, amps, and watts

Item	How to calculate	Analogous to	Works like
Volts	Amps × ohms	Pressure	Pounds per square inch
Amps	Volts/ohms	Flow	Gallons a minute
Ohms	Volts/amps	Resistance	Incline on your treadmill
Watts	Volts × amps	Power	Engine horsepower

Voltage: Hazardous?

Volts give the ability to shock or hurt you, but it depends on how many amps the source can make whether you do get hurt. A 12-V car battery will never ever shock you unless you connect the two terminals to your tongue. Not enough volts. Volts are ability to break through insulation, like your skin. I don't recommend this, but you can lick a 9-V battery and the worst that will happen is that you will see some lights in your brain. I know because I did that when I was a kid. So you can touch a 12 V battery with your hands, and nothing will happen. At the same time, if you get your metal bracelet or wristwatch band across a car battery, you could be seriously burned, because it doesn't have the resistance that you do, and lots of power would flow through it and heat it to the melting point. If you had 1000 amps going through that bracelet at 12 V, that's 12 kW concentrated in that small thing, enough to make it get hot quick and burn you badly—but not shock you.

On the other hand, 120 V from the wall socket *certainly will* hurt you and could kill you, *if* there's a complete circuit from one wire to the other through your body. If you have one hand in your pocket and touch 120 V with the tip of a finger, it still can kill you if you're standing in water in the bathtub, because one side of our power system is connected to Earth. Lots of people have died because of that. That is mostly because the early power companies could save one wire that way, using the Earth for the other.

The reason it can kill you is that your muscles work from electrical signals your brain sends out, so if you get hold of high voltage you'll find you can't turn loose. In that case, the best thing is to remember to fall—if you can.

On the other extreme, 20,000 V from a car ignition will hurt like hell, but not kill you because the source can only make a few thousandths of an amp and ultimately it's current that kills you.

Finally, it's easy to fry instantly in a lightning bolt of power big enough to supply part of a city if you spray paint a 180,000 V powerline in a major substation while you're standing on the ground, as somebody did in Santa Fe not too long ago. That high fence is there to keep you, tagger, out!

Resistance

So what's resistance, then? Think of it as the size of the hole the water squirts through. High resistance requires more voltage to give the same current (amps). It's the way you control the current. In power transmission, you don't like it, but for things like toasters and electric heaters, you do! In a heater, you're deliberately turning electricity into heat.

Power

It doesn't matter if it's 5 V and 1000 amps or 1000 V and 5 amps—it's the same power. But whether it's useful to you depends on the voltage. A kilovolt would destroy your iPad. These days, cleverly designed power supplies can get the power you need for your laptop at the voltage it needs, from 220 V in the wall socket, or 24 V on the airplane without using a transformer.

A horsepower is 746 W. Why that instead of a nice round 1000? James Watt is responsible. He chose a power that would be quite a bit more than an actual horse could do so people would feel happy about his cranky and smelly steam engine, and that had nothing to do with volts and amps in his day. Even metric system countries use horsepower.

Transformers

More often than not, you find a transformer changing voltages for you, like the one on the pole outside your house that starts with 2400 V and changes it down to what is safe for your home (Fig. 2). With the right wiring, the two 120 V circuits add to 240 V for your dryer.

When you go to Europe, you'll find 220 V rather than 120. Why can't people agree on it? On the one side, it's a safety issue. It's harder to electrocute yourself with 120 than 220, and people do sometimes get hold of bare wires. On the other hand, there's the practical issue that switches and wires can be smaller at 220 than

Fig. 2 Transformer outside your house. More than likely, 2400 V is coming in and two 120 V circuits going out (Glogger. GNU Free Documentation license)

120. Why? If something needs a certain number of watts to run, the wires going to it can be smaller if the amps are lower and the volts are higher. Wires can only carry so many amps before they get dangerously hot, but if I need less amps, I can use smaller wires. That's why extension cords and light switches in Europe look so flimsy or, from the other point of view, why our extension cords and switches look so fat and clumsy to Europeans.

In the USA, many shops and small factories use 240 V. Inside big industrial plants, people often use 480 V when they need to run hundred-horsepower motors. So you can see why you might need transformers.

Now, let's say I want to hook whole cities miles apart to a single power plant. A city will need tens of thousands of amps. To do it at 120 V would take a wire as big as your leg, lots of power would be lost on the way and it would cost way too much. Better to do it at 120,000 V or so and get by with a little wire that needs to carry only 10 amps, then transform down to 120 V at the delivery end, to deliver 10,000 amps to all those homes. You see them on tall poles at street corners (Fig. 2). And you see the big ones behind a fence at a central distribution station, like the one that guy tried to tag (Fig. 3).

This is the issue that killed direct current (DC) for power distribution. Today you can transform it, with semiconductor switches, but over the long history of building

Fig. 3 Substation transformer (Sturmovik, GNU Free Documentation license, Wikimedia Commons)

the power grid you could not. If it was DC, it meant your power station had to be close to your homes and factories because all the power had to go at the same voltage at which it was created.

Back in the day, when things weren't yet so decided, Tom Edison, Nikola Tesla and George Westinghouse had a fight over whether power should be sent as DC or alternating (AC). They did competing experiments over which was the best for electrocuting people on death row. Edison liked DC.

Edison electrocuted an elephant. The poor thing wouldn't cross the bridge to the site they had set up for this, so they had to do it on the spot (Fig. 4). Today, we use AC.

Tesla and the First Wireless

Speaking of the Serbian scientist Tesla, no discussion of electricity would be complete without a mention of him and his work. You'll see a bit more about him (Fig. 11 in the section on motors), but the figure most people remember is the man in Fig. 5. He worked for Edison as well as Westinghouse for a while, and garnered $216k in a single payment for his patents from the latter. But what he really wanted to do was to retreat into his laboratory and invent. He used the money to set up his own laboratory in Colorado Springs and play with X rays, radio, a 5000 horsepower bladeless turbine, electric generators for Niagara Falls and schemes for transmitting huge amounts of electrical power over great distance wirelessly. And, of course, the Tesla Coil. I expect you've heard of it. Maybe you've built one? As time went on, his

Fig. 4 Topsy the Elephant at her execution by direct current (1902) (Public Domain)

Fig. 5 Nikola Tesla in his lab in Colorado Springs, 1899 (Public domain)

claims became more and more surrealistic, including thought cameras, death rays—the "teleforce" weapon—and huge mechanical oscillators which were supposed to crack the Earth's crust and destroy civilization. As late as 1937, he was working on particle beam weapons, which also figured in the Strategic Defense Initiative. However, his earlier work caused the metric unit of magnetic field strength (Tesla, 10 kG in the old units) and the electric car to be named in his honor.

Birds on a Wire

You'll often see a flock of birds sitting on the 2400 V wire coming into that transformer. How can they do that (Fig. 6)? For electricity to flow, there has to be a *circuit*. You in your bathtub with a toaster is good enough, because the water is connected to Earth. For the birds, as long as they're sitting on *one wire*, no problem. The electricity can't complete a circuit through them because the air the birds are sitting in has tremendous resistance. Their wire might get a little warm, and that's a good thing for them in winter.

Fig. 6 Birds on a wire (Tomascastelazo. Creative Commons license, Wikimedia Commons)

Different Hazard If It's High Frequency

One time, I talked to a powerline repairman who said he worked on the 180 kV line that feeds Las Vegas, NM with his bare hands, in an insulated bucket of course. I was surprised because I thought you'd have to wear heavy insulated gloves on that job. I imagined he would feel a real shock at 180 kV, just from his *capacity*. Capacity means that electrons momentarily flow into things to charge them up even when there is no complete circuit. When you deal with AC, those electrons are flowing into and out of you 60 times a second. He said he felt "only a tingle."

Then I did the math and confirmed that, 20 m above the Earth in his insulated bucket, his capacity was so small that he would feel only tens of μA up there even at 60 Hz.

It's different with radio frequencies because that capacity current is proportional to frequency, and radio frequencies are tens to hundreds of MHz. Years ago, another friend of mine was building a ham radio transmitter for the 10-m band. Then, those electrons are going back and forth at 30 MHz instead of 60 Hz. Inside the transmitter, the final amplifier was operating at 300 V. My friend was only a couple meters above the Earth (and so had 10 times the capacity of that powerline guy). Then, when he accidentally touched the tuning coil, he probably drew half an amp. Fortunately, he was painfully burned rather than killed. Just so you know ....

AC and DC

Direct current (DC) is the kind that comes out of a battery, always the same, plus on one wire and minus on the other. Alternating current (AC) is a wave that vibrates back and forth from negative to positive and back again, 60 times a second for power distribution in the USA, and 50 Hz in some other countries. Why use AC? A major reason is our need for transformers (Fig. 7).

Transformers

The transformer is an amazing invention due to Mike Faraday in 1831. He was a brilliant, self-taught scientist who did a lot of other things too, like inventing the first electrical generator, but we'll get to that later. Transformers do not work with DC. The reason is they work from *changing* magnetic fields in that core. If you put DC on the primary, you'll get one pulse out and then the core would be completely magnetized. It won't take any more. You have to reverse the voltage to get another pulse. And so on. Do it 60 times a second and it can be pretty efficient.

A transformer will always have a magnetic core, an input (primary) and an output (secondary) coil of copper wire. The ratio of input and output voltages is the same as the ratio of the number of turns.

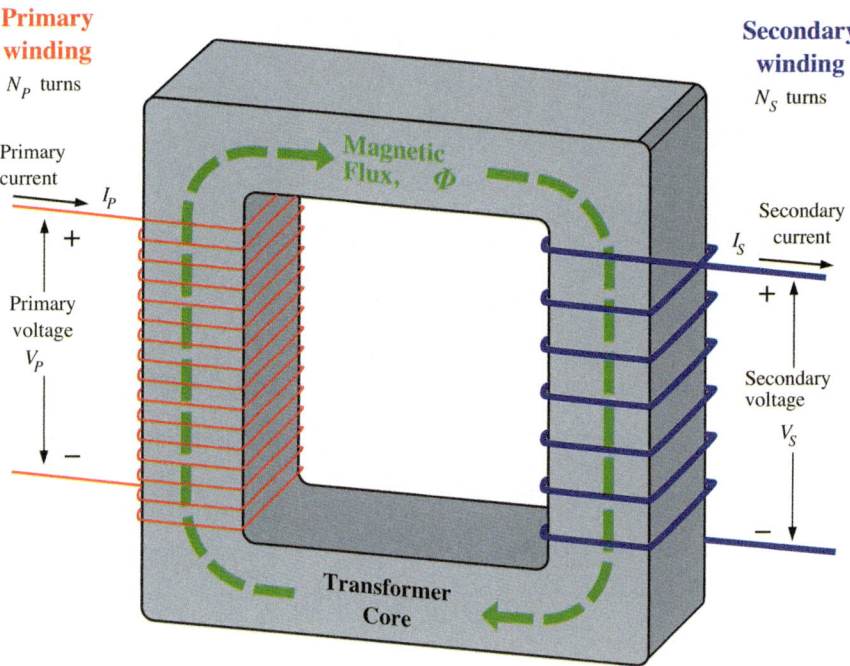

Fig. 7 How a transformer works (Bill C. Creative Commons license, Wikimedia Commons)

Experiment

You can do this experiment yourself! Get a ferrite ring, an NE-2 neon bulb and several meters of small diameter magnet wire with varnish on it for insulation. Twenty gauge is perfect. Wrap the wire round and round through the ring until there are 20 turns. Now, wrap 100 turns of wire in another coil on the other side of the ring. Get a 12-V battery (remember, it won't shock you!) Measure the battery voltage if you have a voltmeter handy. Twelve volts, right? Be sure to scrape the varnish off the ends of the wires with a pocket knife. Wire the neon bulb across the 100 turn secondary. Connect one side of the battery to one end of the 20 turn primary. Connect a clip lead to the other side of the battery and just drag it across the other end of the primary, making sparky, momentary connections. You'll see the little NE-2 light up just like in the photo (Fig. 8).

Congratulations! Because these bulbs won't fire below about 100 V, you have transformed 12 V up to 100 V! I do not recommend making your little sister hold those wires!

Fig. 8 A transformer you can make in 20 min lights a neon bulb! This ferrite ring is 85 mm diameter by 15 mm thick (C. Phipps)

How Do You Use Electricity?

In motors and for light, mainly. Everything from streetcars to your food mixer to the TGV uses electric motors. And if you've ever seen that picture of the Earth from space at night you can imagine how many GW [about 100, just in the USA!] we use just to keep the streets lit. I'm going to focus on motors and generators, not light [see volume 2!].

DC Motors

How does a DC motor work? We can branch out from there to everything else (Fig. 9). Those little long dash/short dash symbols are batteries, and the rods are iron or ferrite. The rotating part is called an armature. All I need to do is arrange it so I have a magnetic field going through space in one direction, and hang the armature in the middle, with switches that go on and off depending on its position. As I'm sure you know, like magnetic poles repel. So, when I put a current i through the armature and give it a magnetic field b, it wants to push away from **B**, and rotates. But then the switch disconnects, waits for it to turn 180° and reconnects, pushing it again. It can go incredibly fast if there's no "load" to drag it down.

Magnetic field? Electric field? See page 6 of *Electromagnetic Waves*. **B** and **E** are the heroes of this chapter.

AC/DC Motors

I said I'd get into AC/DC. One variation (Fig. 10) is called a series motor—instead of having three different power supplies, the whole thing is wired up *in series*. As you can see, it doesn't matter which way the current goes, its still works. So … you can run it from AC or DC.

How Do You Use Electricity?

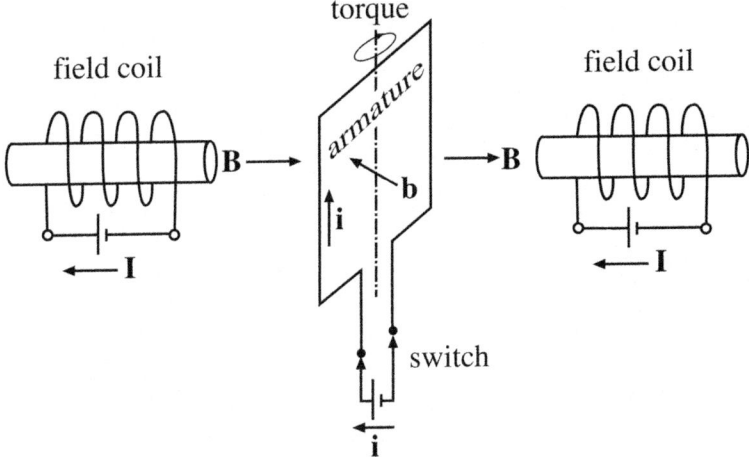

Fig. 9 Principle of a DC motor (C. Phipps)

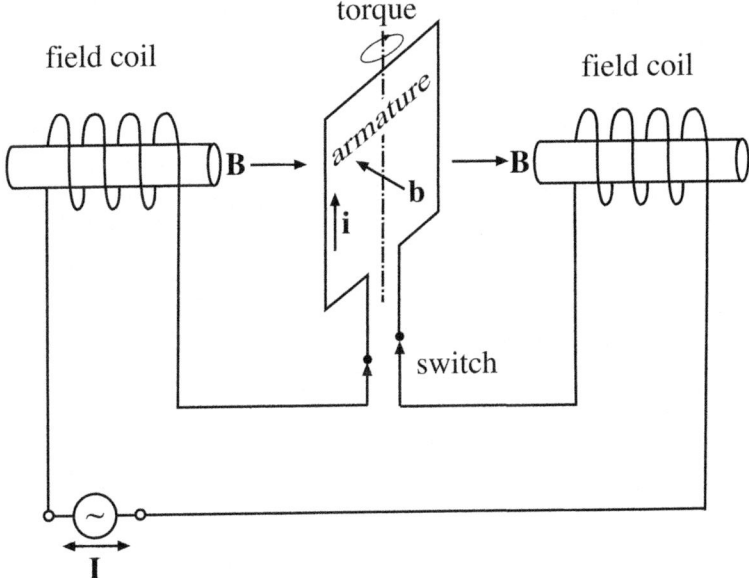

Fig. 10 Series DC motor (C. Phipps)

AC Motors

AC Synchronous Motor
For motors, there are lots of variations. The rotating field b can be a permanent magnet and I can pump AC into the coils that make **B**, spin the rotating part up, and I have a synchronous motor that always rotates at 3600 rpm (revolutions per minute)

Fig. 11 Tesla's induction motor (1887) (Ctac, Wikimedia Commons)

[60 Hz]. It's called synchronous because it always keeps time with the hum of the AC power, and that's one reason you used to see them in electric clocks.

Induction Motor

Or I can short the armature—this was Tesla's genius realization—and make the rotating part a transformer secondary, so you don't even need a current source for it, or a switch, or a magnet! That's called an induction motor, the type most used today. Figure 11 shows the actual motor he built.

You need to spin induction motors up to get them going, but it accelerates much better than the synchronous motor and puts out a lot more torque. *Torque?* Ah, yes—that's just rotary leverage, force times length of the lever.

How Do You Make Electricity?

Batteries

Of course, the first way was batteries. Alessandro Volta made the first one about 1800, unless those tricky Arabs did it first 2000 years ago as part of an electroplating setup. He used a "pile" (the French still call batteries *piles* if you are in Paris and need one!) of alternating copper and zinc disks separated by layers of cloth or cardboard soaked in salt water. Salt water makes the zinc want to lose electrons more fervently than the copper so that you get an electric current going through this stack of metal disks from the last zinc to the first copper disk. The energy comes from a chemical reaction. Volta could explain why Galvani's froglegs jumped using this device. You can read about that.

We still depend on batteries, a lot. Lead-acid, "alkaline" (manganese dioxide and powdered zinc), lithium-ion, nickel-cadmium and nickel-hydrogen are the main examples. You probably use one or more of these a lot if you drive a car, use an iPhone or drive a Prius or Tesla.

Those with the highest energy density (lithium-ion) can store up to 1 MJ/kg. This sounds like a lot until you realize *gasoline stores 42 MJ/kg*. And, as we all know, lithium batteries can explode. That's because they're pressurized *and* have a flammable electrolyte. The main sources of lithium are in Bolivia and Chile. Contrary to what you read, we won't actually run out soon—there's enough to cover today's usage for 400 years. But it is still a precious resource.

Solar Cells

Sunlight hitting a slab of two types of silicon with different impurities in it will make an electric current. This is very attractive at first blush: there are no moving parts and all it needs is sunlight, of which there is a lot in the American deserts and elsewhere. No carbon footprint. Of course, making silicon in the first place has a big carbon footprint. And, power from solar cells today is more than four times as expensive as power from windmills, and about quite a bit more expensive than the coal-fired plants we all love to hate. But—that it's even in the same ballpark is an amazing engineering hero story in itself.

Generators

By far the most important way we make electricity is with rotating machinery, whether driven by the wind, a gasoline or diesel engine, a nuclear reactor or a waterwheel. Of these, believe it or not, power from wind machines costs 40 % less than from those coal-fired plants.

Which brings us to: how does a generator work? And that's one of the main reasons I got into this topic. It's just the reverse of "how does an electric motor work?" so it's interesting from two points of view. You can buy a kit for $9.95 that shows you both motor and generator action.

For a DC *generator*, I can rotate the armature in the field (Fig. 9), throw away its battery, and take out the current where the battery was. My effort cranking it goes into making power.

Or if the armature is a permanent magnet that I rotate, I'll get AC out of the field coils, and that's a generator. In fact, they call it an alternator, and you have one in your car. So that finishes ordinary motors and generators.

Faraday's Homopolar Machine

I apologize for the previous section being a bit encyclopedic, and therefore more boring than usual. Now, it is my pleasure to introduce you to one of the most fascinating, death-defying electrical machines in history, the Faraday Homopolar Machine. It was actually the very first generator back in 1831, and was invented by the genius Mike Faraday. It has nothing to do with gay people. It is just a disk on a shaft turning in a crosswise magnetic field, and it can be a motor or a generator (Fig. 12). I'm showing it as a generator.

So far, so good. Even if it is a weird geometry, you move a conductor through a B field and you get a voltage, right?

Now, look at Fig. 13!

In any normal generator, the field is stationary and the armature rotates, or vice versa. Here, the *whole thing* rotates, field and armature, and it is still a generator! How can that be? The answer is that what does rotate with respect to the fields is the stream of electrons in that armature disk as they travel out to the *stationary* brushes that pick off the current!

Experiment

Needless to say, I built one! You can, too (Fig. 14). What you see in the lower center between the supports are a couple of large coins, in this case old Greek 100 drachma pieces, sandwiched between two *really* powerful fridge magnets that make 1.3 T

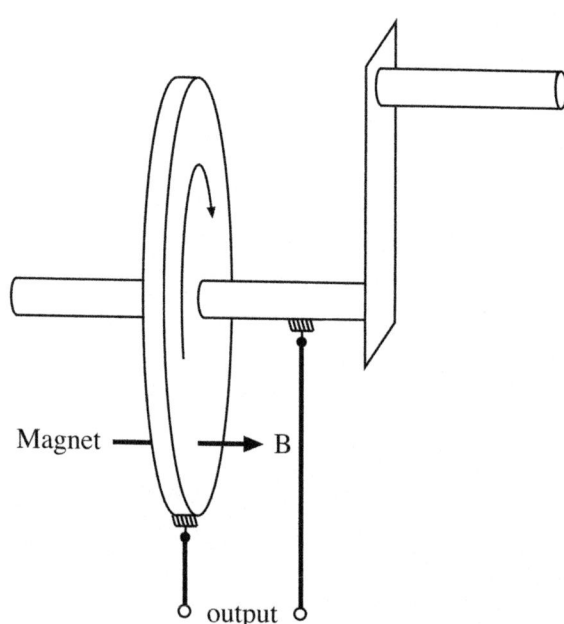

Fig. 12 Faraday's Homopolar Machine (1831) (C. Phipps)

How Do You Make Electricity?

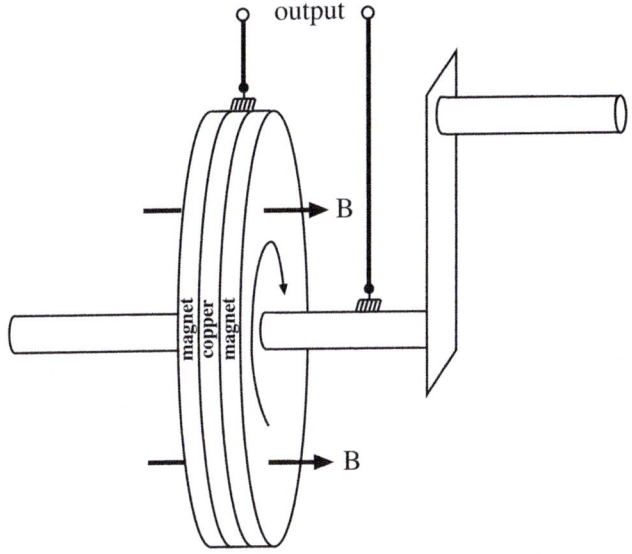

Fig. 13 A mindtwisting version of Faraday's Homopolar Machine (C. Phipps)

Fig. 14 A real Homopolar Machine (C. Phipps)

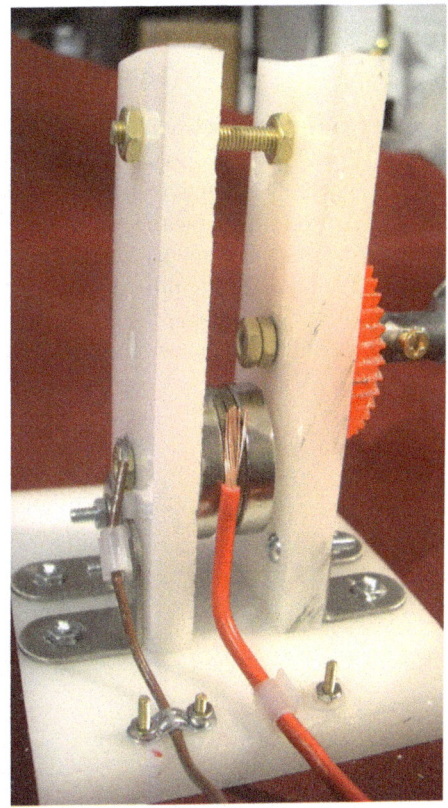

(13,000 G in the old units). The **B** field goes from right to left through the thickness of those disks, parallel to the shaft. In the photo, you'll see the chrome plated edges of the magnets on both sides of the coins, which are just under the brush with the red insulation. The other brush touches the axle on the left. Thanks to the red gear on the right, I could get the rotating sandwich up to 1800 rpm.

This Faraday Machine made just 40 mV, not even enough to light an LED. This is what I expected, even with 1.3 T, because the length of the path along the radius of the coins from the shaft to the edge is only 1.5 cm. I am not strong enough to assemble anything larger with this design. The force between the magnets was 39 pounds! You could get a lot more volts out of, say, a 30-cm diameter disk with these magnets all around the edges, but I couldn't afford that option.

Assembling this version involved lots of swearing, Our cleaning lady was afraid to come upstairs to the Lab. Every magnetic object within a foot pointed in its direction. All my tools are now magnetized. *But* it *worked*! I had to build it because it's so hard to believe that it does.

By the way the magnetic pressure (see *Weird Reality*) created by these two fridge magnets is 2.5 times what you have in your tires. You can imagine they're dangerous if they shatter, so be careful. Don't try to use pliers to assemble them!

What's it good for beside being a fascinating toy? Large Faraday Machines are particularly good for making single electrical pulses of very high energy (tens of MJ, MA) as they dump rotating kinetic energy into a dead short in a few ms. They can use high temperature super-conducting fields and liquid metal brushes.

Electromagnetic Tethers

Here's something else that works in space and can be a motor or generator, (Fig. 15): a satellite traveling East with a big wire hanging down in the Earth's magnetic field. Put an electron emitter on the end and you'll get kV as you zoom through the Earth's field at 7 km/s! The circuit is completed through space. Of course nothing is free and if you use power from the tether, you're gradually using up the kinetic energy of your satellite. Or you can put out a current from your solar array and raise your satellite!

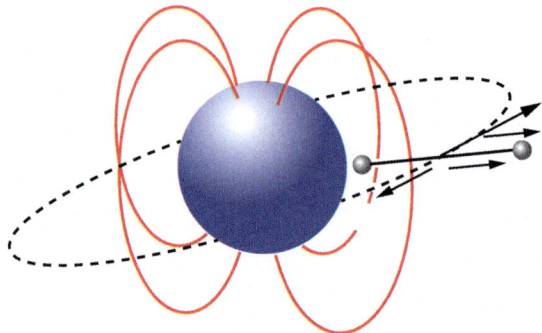

Fig. 15 A real Homopolar Machine (NASA Public Domain)

The Earth Dynamo

But what *makes* that magnetic field that the tether drags through? *The Earth is a giant dynamo*! But we don't understand it. What we do know is that the Earth interior is molten metal until you get to the solid core, and the liquid metal moves. Moving metal by itself won't make a current because you need a moving positive or negative charge to do that and the metal would short that out. If you set up some current in the Earth, you can calculate that it would dissipate in 15,000 years of its own accord. So, "it's a self-excited dynamo depending on a pre-existing field." But what made that? People have done all sorts of really detailed modeling of the *flow*, but what makes the current? Another explanation is that the solid core is a big magnet. But what magnetized it? There's something for you to work on!

Lightning!

Dozens of times a second, lightning strikes somewhere on Earth. We're talking 50 MV to drive this thing through 5 km of air and 100 kA peak currents during 50 μs or so. The total energy can be 250 MJ for one of these! That's a lot of energy and it makes a lot of sound. The lightning bolt itself is just 1 cm in diameter despite how big and bright it seems, and the temperature in there can be 300,000°! First, there's a downward stroke and then a much brighter upward-going one that can go a few percent of the speed of light.

Now here's the interesting point of Fig. 16: thunder comes from all the kinks in a lightning stroke! If lightning were one straight cylinder, acoustic theory says you would just hear a "swish."

Fig. 16 Lightning and thunder (F. Wicke)

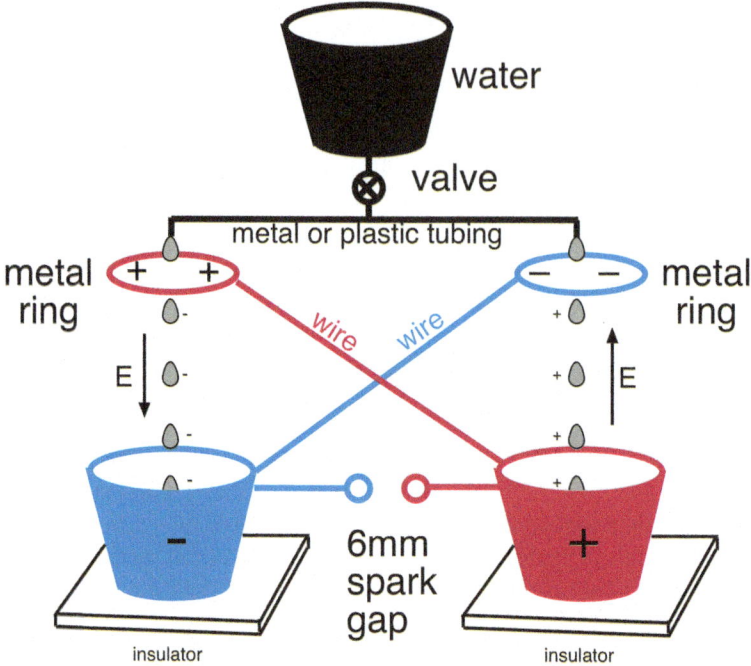

Fig. 17 Kelvin water-dropper (By C. Phipps, own work, based on Qwertyuiop GNU free documentation license vis Wikimedia Commons)

Why does lightning happen? Here's another thing we don't really understand. Most references say that somehow charge differences build up in a cloud. The energy comes from the huge vertical drafts in a thunderstorm, which is anyway like an atomic bomb going off, so there's plenty of energy. Figure 17 shows my bet for how it happens.

Experiment: Liquid Van de Graaf

This is an experiment you can do yourself, due to another great scientist, Bill Thompson, otherwise known as "Lord Kelvin," for whom degrees K were named. I saw this working in a Tulsa science fair, and I was very impressed. This will twist your mind! All you have to do is make sure the insulators under the pots are *really good*, that the wires are a few cm apart and that the spark gap is hanging in air and not touching anything.

At the start, both pots have the same electric potential. Now you start the drips. Somehow, one drop (say the one headed for the blue pot) has one more electron in it than the drop on the right. It makes the blue pot slightly negative, and charges the blue ring negative. That will *attract* droplets with a little imbalance toward *positive* charge into the red pot. Now the red ring is positive, and that attracts droplets with an imbalance toward *negative* charge to the dripper over the blue bucket. The next drop brings two positive charges to the red pot because of the electric field that's already there. Then 4, then 8, then 16 ... as time goes on, can you imagine this is another *Exponential* process? Gravity is powering this. Eventually, *really positive drops are falling through a strong electric field that repels* them going from the blue ring to the red pot, and they have to do work to get there. Really negative drops are doing the same thing on the left hand side of your setup. At the end, the field rises to the kV level and ... PAF! You'll get a really satisfying spark! Be careful not to touch it. Do you see how columns of droplets being blown upward and falling for km in a thunderstorm could make MV between clouds, and to the ground?

Sprites

Here is yet another thing we don't understand. Many km above a thunderstorm, in the ionosphere, here is what you can sometimes see (Fig. 18). All those megavolts in a thunderstorm do something in space, too! Isn't that beautiful? This one is a

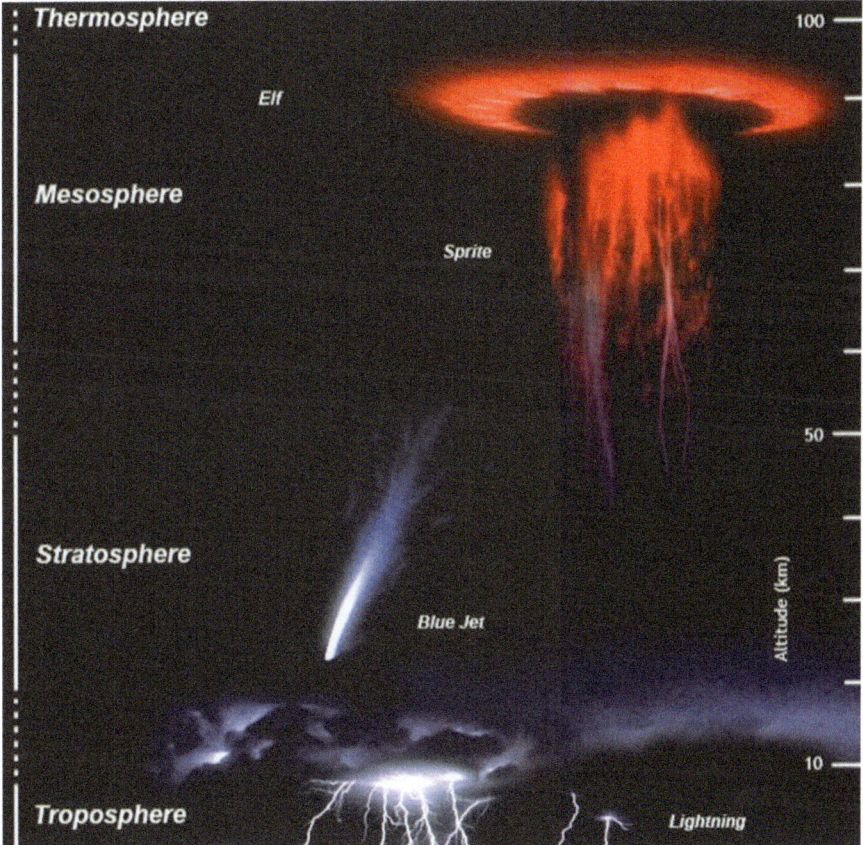

Fig. 18 Sprites (Abestrobi Creative Commons license, Wikimedia Commons)

"jellyfish sprite," for obvious reasons. There are also carrot sprites and column sprites.

You see them from aircraft and you even see them from space! Far away from a thunderstorm on the ground you can also see and photograph them.

Sprites send out X ray and gamma ray bursts.

Wimshurst Machines and Capacitors

This wonderful object (Fig. 19), familiar as a source of terror for lots of helpless girls in Frankenstein movies, is an electrostatic generator. A process like sliding slick shoesoles on wool carpet in winter builds up an electric charge which is stored in "Leyden Jars," which are just simple high voltage capacitors made from glass jars with tin foil on the inside and outside.

Capacitors? I mentioned capacity earlier in this chapter. A capacitor is something that has electrical capacity. Just another way to store electricity (Fig. 20). You've probably seen them if you ever looked inside anything electronic. They're different from batteries, which also store electricity, because they can dump it very fast, even in billionths of a second for those little brown pillshaped ones in the figure. Leyden Jars are romantic, but nobody uses them anymore. You can make your own Wimshurst Machine with a CD. Check out the internet for instructions!

Fig. 19 Wimshurst machine and two Leyden Jars (Wikipedia public domain)

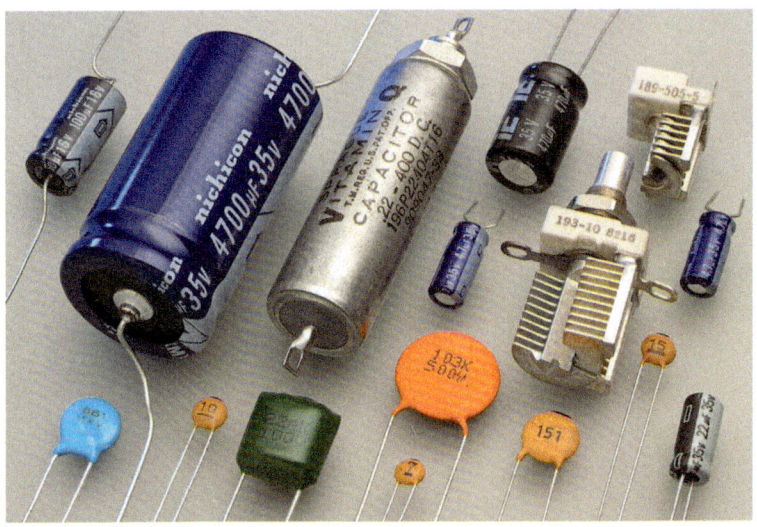

Fig. 20 Capacitors (Eric Schrader. Creative Commons Attribution-ShareAlike 2.0)

Conclusion

Volts, amps, ohms, and watts are all different, and it's important to know the difference just to be able to vote on things intelligently. You won't learn about these things in the news. There are lots of really interesting kinds of motors and generators, including the Earth itself, and lots of safe and fun experiments you can do if you want. Lightning is one of the most interesting examples of electricity flowing. Sprites are fascinating too, and you might want to observe them yourself.

Next, let's go on with the development of drones, jet planes and rockets!

Drones and Robotic Flight

No pilot you can see, but he can see you (Fig. 1). That is a remotely controlled drone. Sometimes controlled from thousands of miles away, but controlled. This is an unmanned aerial vehicle (UAV), different from a robotic craft.

As a military craft, the concept goes way back—at least to the mid-1800s when Austria sent balloons over Venice to deliver bombs.

At least 28 countries use UCAV's (Unmanned Combat Aerial Vehicles). The attraction to users is clear: I don't have to put my men and women in danger. Sometimes the operators are in Nevada and the targets in Iraq. But there are political side-effects not many folks think about. One is "blowback," which just means unintended consequences, perhaps achieving the opposite of what we wish to achieve, making many more enemies rather than less. Not only that, local governments which permit other nations to use drones over their territory may sacrifice their legitimacy with their own people.

You can use the same type of craft for peaceful purposes—search and rescue and powerline surveillance (Fig. 2), for example. Wankel engines have been used in UAV's.

The mind boggles. What about spraying crops, exploring for oil and gas, protecting archeological sites from vandalism and tracking schools of whales? Cars are made by computer-controlled robots, and laparoscopic surgery is done that way. You must have heard of the Roomba vacuum cleaner.

There are even robot teachers. I bet you think that's what some of yours are!

And, as we've heard recently, Amazon may start delivering things with drones. It's going to be called "Amazon Prime Air." They're testing them in Canada right now to avoid hassles in the USA.

The regulatory folks haven't yet figured out how to deal with drones.

Fig. 1 A (rather grim) Reaper (Sgt Brian Ferguson USAF public domain)

Fig. 2 Surveillance Drone (CSIRO Creative Commons License via Wikimedia Commons)

Robotic Warriors

If a scientific mission is so complex that isn't practical to control it remotely, a robotic research vehicle is the answer. If it's so complex and dangerous that men can't do it—like clearing radioactive rods from a reactor explosion—then, again a robot is the only solution. From that to deciding ground wars need to be fought with robots is only a small conceptual step once the technology is good enough. What kind of political and social problems might arise from that decision? What about swarms of robots (Fig. 3)? These are not military robots, but they could be. Think about it. In the 2003 Iraq War we had semi-autonomous cruise missiles.

Samantha was mentally and socially anthropomorphic, just not physically so. It is unlikely that a military robot swarm would consist of physically anthropomorphic units, but it could.

We continue the topic of autonomous drones and robots in greater depth in the Chapter on *Electronics and Computers*.

Fig. 3 A robot swarm (Sergey Kornienko GNU Free Documentation license via Wikimedia Commons)

Conclusions

Drones and robots offer tremendous advantages for lots of things humans don't want to do. At first sight, there are no problems. Then, when you think about it a bit, especially from the other person's point of view if these are military robots, things get more complicated.

The Jet Plane: How Metal Birds Fly

I think when they dig up our civilization in 10,000 years, whatever "they" are will really be impressed that we valued visiting each other enough to create the jet plane. Traveling at almost the speed of sound (or even faster), this thing can take us anywhere on Earth and, on the flight from New York to Jo'burg, do it without refueling in less than a day. The jet plane is one of the most amazing creations of the twentieth century, but most of us take it for granted. The Heinkel 178 was the first (Fig. 1).

In this book, I usually avoid history. But the history of jet planes is too fascinating to ignore.

As early as 1910, a Romanian engineer named Henri Coanda began imagining a whole new kind of plane. Although he had the concept, there is no evidence that it could fly. As is often the case, the French had the first practical idea, and the Germans built it first. Maxime Guillaume patented the gas turbine in 1921. In Germany (Fig. 2) Hans von Ohain of the University of Göttingen patented his idea for a jet plane in 1936, made the first operational engine in 1937, and, with the Heinkel group, the first flight in August 1939, flying what was labeled a Heinkel He178. Amazingly, a movie of that very first jet flight 74 years ago still exists. Although it could only stay up for 10 min, about as long as the Wright Brothers' first flights, it reached what was then the astonishing speed of 700 km/h [435 mph].

Like many German scientists, Ohain was brought to the USA after the war under *Operation Paperclip*, working first at Edwards Air Force Base in California, and, instead of being hung, rose through the ranks to Director of the Wright Patterson Aeronautical Research Laboratory and finally became Chief Scientist. There, he worked on things like nuclear-powered and vertical takeoff jets in the 1960's.

It pays to know stuff! He lived until 1998.

Frank Whittle independently developed the concept of jet engines in the UK, and patented his idea in 1930. However, his first operational turbojet was not flown for more than a few seconds until January, 1941.

In developing a real, functioning jet plane the Germans beat the UK by 9 months in April, 1944 with the first operational jet fighter. This was the Messerschmitt Me 262, which went 150 km/h faster than the fastest Allied fighter, the P51 Mustang.

Fig. 1 The He178 (Public Domain)

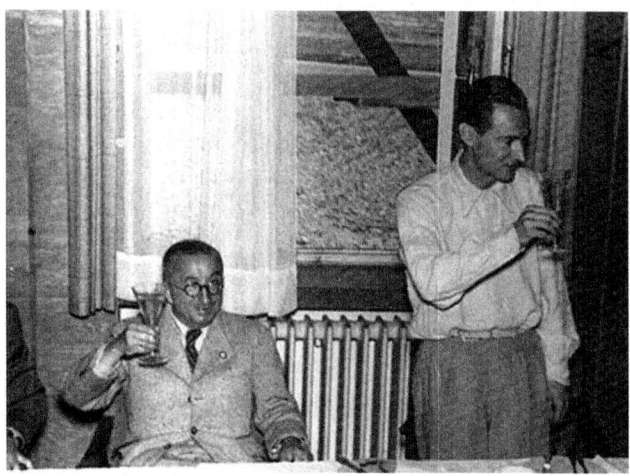

Fig. 2 von Ohain (L) and Heinkel celebrating the first jet flight in 1939 (Source unknown. Public domain)

But then it was too late. While more than 1400 Me 262s were built, only 100 were brought into service. Oddly enough, these early jet engines used centrifugal blowers rather than turbines. By 1944, Anselm Franz of the Austrian Junkers company had figured out the axial flow turbojet, operating on the same principles we use today.

First Commercial Flight: The Comet

I still remember reading a breathless Popular Mechanics article in 1952 entitled "I flew across the English Channel at 550 mph!" It was about the British Comet, the first production jet airliner. The guy who wrote that article was very lucky, because these planes soon began experiencing a lot of fatal crashes. Queen Elizabeth, the Queen Mother and Princess Margaret were also very lucky when they flew on it in 1953. Just like today, cabin pressure was kept at a comfortable level for passengers even though the outside pressure is much below that on Mt. Everest, and you would asphyxiate quickly if you were out there. So, each time the plane takes off, it has to

Fig. 3 The deHavilland Comet wing (Ian Dunster, Creative Commons Attribution-ShareAlike 2.0 UK: England & Wales)

be pressurized and the metal skin and struts stretch. Have you ever bent a piece of metal back and forth until it breaks? That's metal fatigue, and that's what got the Comet, which quit operating in that version by 1954. It did fly to Jo'burg and back though. And London to Tokyo was just 35 h instead of 85.

In my opinion, its sleek wing design is one of the most beautiful things that have flown (Fig. 3). Do you see what's different? The engines were buried in the wings instead of hanging down below them like today. The design made it much harder for a bird to fly into the engine.

But, by 1953, the wings literally started falling off. As the number of flights for each plane added up, they literally began exploding in flight over the ocean, with no survivors to explain what happened. There were thirteen fatal crashes in a couple of years. They found out what was wrong by building a huge tank in which water pressure could be used to repeatedly pressurize a whole plane. One of the main problems turned out to have been square windows, which concentrated stress at the corners, and the wrong kind of rivets. The plane was restored to service in 1958. After all, it was the very first commercial jet, and failures were to be expected, but they were not diagnosed and fixed in time to save the plane. By then there were lots of other competitors who had not had problems, and it was too late for Comet.

It's not talked about much these days, but some of you may remember the same kind of thing happening on early flights of the Lockheed Electra. These were the first commercial airline planes with turboprop engines. One of those, and its passengers, was found at the bottom of a 20-meter-deep hole in Tell City, Indiana in 1960. It had been Northwest Orient flight 710 before the wings fell off. They had straight instead of swept-back wings and engines mounted above the wings. There was a failure mode that came from an instability (see *Exponentials*). The engine would pitch down a little bit, changing thrust, then pitch up a little more, then down even more until it twisted the wing off. Since nobody lived to tell about it, it was a long time before they figured out what happened. Then they just strengthened the wing.

Moral of this story? New technology will have unforeseen problems. You can't test too much if people are involved. These days we use supercomputers, but they didn't have those then, and they could only test what they could think of.

Jet planes can go even faster than the speed of sound, and we will talk about that later. But first, let's take a jet plane apart.

How Can a Metal Bird Fly?

This is an interesting question, which a friend who read the first version of this chapter immediately asked.

It all has to do with power-to-weight ratio. What's that? It's just the power output of the animal or plane divided by the weight. And the weight is an interesting story, the result of making a light alloy.

Birds can only put out from 10 to 25 W/kg of body weight. Even the strongest eagle.

Of course, humans are much weaker in flight, at about 3 W/kg, eight times less than the strongest bird. That's easy to see if you imagine doing an "iron cross" gym exercise and repeating it twice a second like an eagle does.

How do we know these numbers? That's a very interesting story.

Those who have tried to fly like birds by pedaling an airplane have a difficult job. It requires top training, plus an ingenious aircraft design to get the weight of the whole structure (man plus plane) down to where it can be done at all and, of course, the ability to pilot a plane. Lance Armstrong can put out nearly 7 W/kg of his own body weight, or about 525 W, because he weighs 75 kg. But he doesn't have wings and is not flying. For human powered flight, wings and a propeller (or wind-pusher of some kind) are needed in addition to the human, and we need to divide those 525 W by the weight of man plus plane to get a number that we can compare to what birds can do. Even the lightest airplane that a man can fly by pedaling (so far) weighs at least 24 kg. And that is a truly amazing accomplishment by itself! Adding his weight to the plane's weight, a trained human like Lance could achieve about 5 W/kg of the whole structure including himself. Of course, this is only for a brief sprint (Fig. 4).

In 1988, Kanellos Kanellopoulos flew the 110 km from Heraklion to Santorini in an MIT-designed plane called Daedalus in honor of the mythical person who first used manned flight about 3500 years ago to escape Crete. That was the man-powered plane that weighed 24 kg. Kanellopoulos was unusual for athletes in meeting the three requirements of power, weight and piloting ability. He pedaled and flew Daedalus for nearly 4 h at an average speed of 30 km/h to a world record in 1988. For this flight, 3 W/kg was the result. That's how we have that number.

Now here's the point of this section: A 767 jet plane weighs 185,000 kg fully loaded at takeoff. But, during climbout at about 380 km/h, its engines can put out an

Fig. 4 Why it's hard for men to fly (F. Wicke)

astonishing 136 million watts [94,000 horsepower] at maximum power output. The ratio here is 733 W/kg, 29 times what the strongest eagle can do and about 240 times what the best of us puny humans can muster.

How is that possible? For one thing, even though it's made of aluminum or, lately, plastic, that huge plane is lighter for its size than a bird. Let's just look at body size, because it's hard to compare wings. I can't find the body volume of an eagle anywhere, but I'm guessing 2 L. It's mass is about 7.5 kg, so 3.75 kg/L. For the 767, the cargo space is 438 cubic meters, or 438,000 L, giving only 0.42 kg/L, *nine times lighter for its size than an eagle.*

And the engine is stronger than the bird's. Put them together and you're bound to get good results!

Wings

The airplane wing is one of the great inventions of the twentieth century, and it makes it possible for machines to fly. The only thing a bird can do that a passenger airplane can't do is to flap its wings. Maybe someday!

The purpose of a wing, whether it belongs to a bird or an airplane, is to make the upward push that balances the weight of the bird or airplane. The upward push is "lift." Because it can't flap, the wing just cruises.

Lift has two parts, both of which are easy to understand.

The first part of the lift force is obvious. When you were younger, did you ever make a fan out of tinkertoys and those flat paper fins that used to come with Tinkertoy® sets? Whether you did or not, you've felt the wind come off a window fan on a hot day (Fig. 5). If the blades are tilted, spinning the fan makes the air move forward toward your face. Of course, when the blades push the air, the air pushes back on the fan.

In the same way, as a tilted wing moves from right to left in Fig. 6, the air is moved downward and pushes on the *bottom* of the wing, lifting it up.

Fig. 5 A fan (C. Phipps)

Fig. 6 Cross section of a wing with normal lift going into stall as angle of attack increases (F. Wicke)

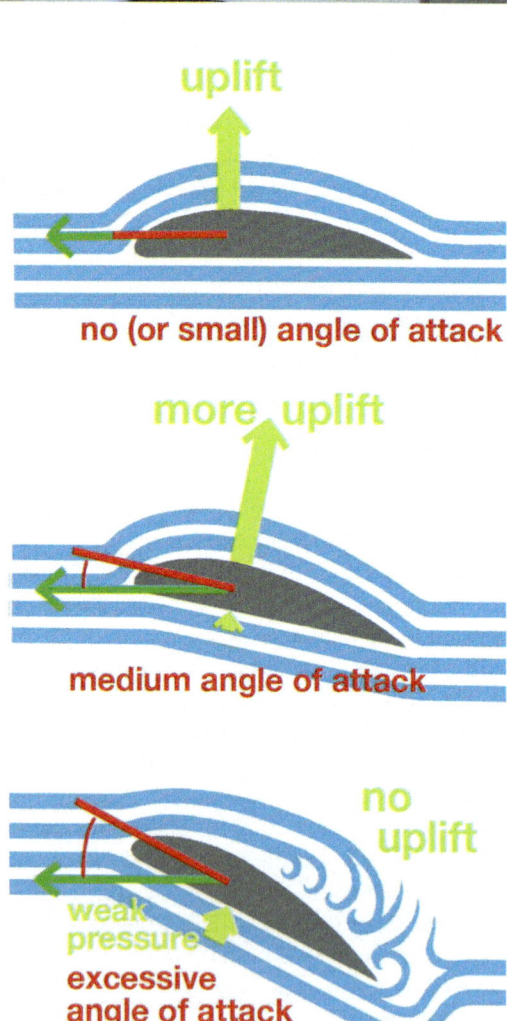

The ingenious part of airplane wing design came when designers studied birds' wings and realized the "cross-section" is not flat but curved on the top side. At low-to-medium angle of attack, as the wing moves from right to left, the air over the top part of the wing whirls around the curved top shape and pulls up on the *top* of the wing. Why? Because the air has farther to go on top, it goes faster than on the bottom. I know it's not intuitive, but faster air has lower pressure. A good pitcher uses the same phenomenon to launch a "curve ball." The spin he puts on the ball makes the air on the bottom move faster than the air on the top as it flies toward the batter, pulling it down.

The two forces together lift the airplane more efficiently than the bottom side force acting alone.

Except when they don't [bottom part of Fig. 6]: if you tilt the wing too much by trying to climb too steeply when the plane is going too slow, you get a situation called "stall" and the plane crashes. That's because the air on the top actually pulls away from the wing so there's no lift. Worse, too much of the lift force is pointing in the wrong direction: backward to slow the plane down, instead of up, to lift it.

Moveable "slats" on the front edge of the wing, and flaps that extend at the back edge can increase the size and curvature of a wing just like a bird does when it's landing or taking off. You've seen that.

Ice on the wing can destroy the lift by messing up the flow of air and make a plane drop like a rock. If you combine that with a stall, as happened on a flight headed for Buffalo, New York in the winter of 2009, you get a fiery crash. The other similar case that comes to mind in many years is Flight 447 from Rio to Paris in 2009.

This happens very rarely, and it's usually pilot error. These days in the USA, 270 times as many people die each year from car accidents as from plane accidents. Put another way, you need to fly two billion miles to have a high chance of dying.

Figures 7 and 8 show other uses for wings.

Intelligent creatures (Fig. 9) rapidly change the shape of their wings while they fly, which makes for much better control and efficiency than machines have achieved.

The Turbojet

Now, what drives a jet plane? Have you wondered how a jet engine works? That's something we take for granted, even though it's a very new invention.

I mean, it's clear most of you boys how the engine in your car works. Fuel and air go into the cylinder, get compressed and ignited, and the explosion pushes a piston, which turns a crank, which turns the wheels. Or take a steam engine. You can see that working, because the cylinder, piston, and crank are all on the outside in plain view [see *Machines*]. The only difference between a steam engine and a car engine is that fuel burns and heats water to make steam, and valves let the steam into the cylinder at the right time to turn the crank, which is also the front wheel.

Fig. 7 Wind turbines (Hans Hillewaert/ CC-BY-SA-3.0)

Fig. 8 A helicopter (Jordi Payà under Creative Commons Attribution-ShareAlike 2.0 Generic)

Fig. 9 Pterodactyl (Dmitry Bogdanov, 2006 by permission under Creative Commons Attribution-ShareAlike 2.0 Generic [see GNU Free Documentation License Version 1.2])

But a jet engine? Here's this thing on the wing. Fuel goes into it, gets ignited and makes a flame. Then, the main idea is pretty clear: hot air squirting out the back end pushes the plane. But, wait. It's not just a pipe with one end closed. It's open on both ends and you can see the fan where air goes in the front to feed air to the flame. Why doesn't flame belch out both ends and make no push at all? Well, you say, it's clear. The fan blows the flame out the back. But, let's see. What turns the fan? A turbine sitting in the flame that goes out the back turns the fan. It sounds just like perpetual motion when you think about it.

I used to have a friend who thought that if he mounted a fan on the roof of his VW van and connected it to a generator, he could use the electricity to run a motor to get better mileage (Fig. 10). Does that seem reasonable to you? It shouldn't. People are always talking about perpetual motion, but nobody's done it yet!

So, here's a picture (Fig. 11). Think about it. You can see which way the shaft has to rotate to turn the compressor so the air comes in, instead of out. In order to work, both fans have their blades tilted the same way, like they are in the picture. I have a fire in the middle. If it went the other direction, it could also turn the shaft in the other direction! How does the shaft know which way to turn? Why doesn't fire sometimes come out the front instead of the back and really surprise the pilot? Or both ends, because the fan just can't decide?

To get to part of the answer, we'll take a quick detour through a hydraulic-ram water pump. You live in the northern California hills and you'd like to be off the grid. There's a stream at the bottom of the hill and you'd like to get the water up to where you live without electricity. You build a hydraulic-ram pump (Fig. 12).

You lay a pipe about fifty feet long in the stream. The little genius invention at the end works like this: when the water gets going fast enough through the pipe, it pushes the blade down and stops the flow. When the flow is suddenly stopped, there's a spike of pressure that drives some of it up the vertical tube, and the check valve keeps it from going down again. After the pressure pulse, because there isn't any flow, the spring rotates the blade up and the water begins flowing again to repeat

Fig. 10 Perpetual motion? (F. Wicke)

Fig. 11 A confused flame in a turbojet engine? (F. Wicke)

the cycle. After several cycles, water starts coming out the top of the hose connected to that vertical pipe, maybe 50 feet above where it started in the stream. Does *this* sound like perpetual motion? In this case, it shouldn't.

So, what does this have to do with jet engines? Figure 13 is a cross section of an actual engine.

Just looking at this tells you a lot. Do you see the 17 big fans lined up one in front of the other at the intake? Each one makes the pressure behind it a little larger, let's

The Turbojet

Fig. 12 Hydraulic ram pump illustrates impulse pressure (F. Wicke)

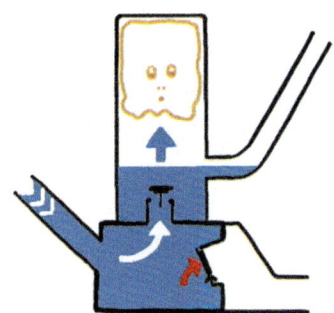

1) Rushing water slams the door at red arrow, makes high air pressure.

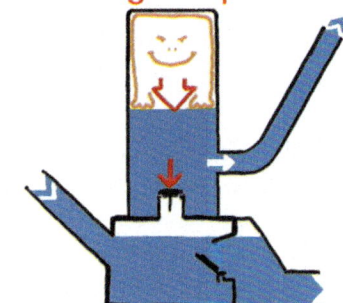

2) Air pressure pushes water uphill, far above the stream!

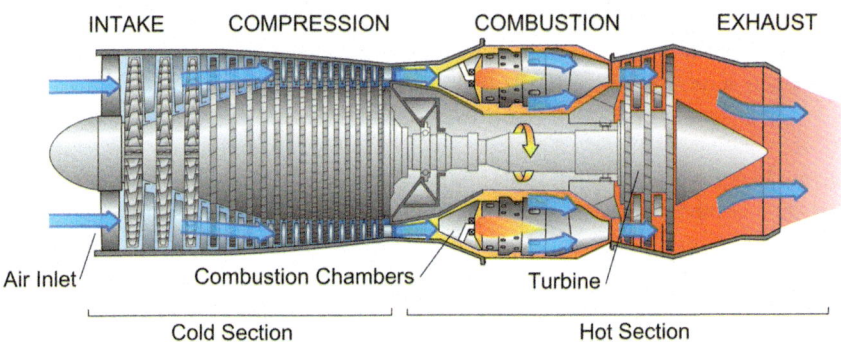

Fig. 13 The turbojet engine (Jeff Dahl under GNU Free Documentation License version 1.2)

say 25 %. But after 17 fans, 1.25 × 1.25 × 1.25 ... 17 times can give a pressure 44 times what it is in the outside atmosphere! This is amazing. It's even a higher compression ratio than in a diesel engine!

But we're still avoiding the main question: why *does* the flame go one way and not the other? And, what does the hydraulic ram pump have to do with this?

Just like in your car, this engine needs to be started. On the runway you've probably heard the whine of one of the turbines being sped up before fuel is allowed in. Imagine that we haven't yet turned on the fuel, but just sped up the turbine to a speed like in a blender. Just at the end of the blue arrow in front of the combustion chamber, the compressed gas hits a wall. Not stopped completely, but slowed down enough that its pressure gets very high like in that water pump. Not only because of the compression but also because of the effect of hitting a wall, just like in the hydraulic ram pump. *Much higher than anywhere else in the combustion chamber.*

Now we add fuel and ignite it (the pointed orange flames in the combustion chamber). The first effect is a "boom" of sound when the fuel explodes in there. You've heard that too before takeoff. The flame is so hot that it would melt all the metal parts if we didn't let cooling air (the blue arrows) flow around the flame and mix with the hot gases before it hits the metal parts. The flame adds heat to the gas, but the place where it goes out of the combustion chamber is much bigger than the place where it came in, resulting in a lower pressure. Because the pressure at the input is very high, while the pressure at the output is lower, even if it is hotter, the gas will always flow in one direction from high to low pressure. Force is pressure times area, so you get "thrust."

Now, it expands into a very large chamber in the red-orange exhaust region and reaches the speed of sound. Because the gas is flowing so fast and because the turbine blades are big, it takes only three, not 17 blades in that section to collect enough power to run the compressor fans and still leave a lot of extra power to push the whole airplane, like it should.

Believe it or not, it's a close call. It takes two thirds of all the power generated by burning the fuel *just to turn the compressor*. Only one third pushes the plane!

That's the whole story of how it works, except for one more innovation, which all airliners have used since 1980 or so, the turbofan engine.

When I saw the first 707 take off from the runway in Boston in the early 1960s, jets used engines that worked pretty much like in Fig. 13, and it took three miles of runway to get off the ground. It was much faster than a propeller plane, but it took *forever* to climb to a cruising altitude. So, people looked for ways to increase the push, called thrust, at takeoff. Here's the result (Fig. 14 — isn't that beautiful?):

The center part is the same as in Fig. 13 but there's a *much* larger fan at the front. This is called the turbofan. Why is it a *turbo* fan instead of just a fan? Like a lot of things in our society, the added word makes it sound sexier. The French call it a turboréacteur. The Germans, a turbojet. This larger fan just blows outside air around the whole engine for even more thrust. It's like a turboprop in a can. Next time you're getting on the plane, look at the engines. You will see that big fan at the front. Computers can tell you how to shape the blades so they're efficient.

Fig. 14 The high-bypass turbofan engine: Rolls-Royce Trent (Science Museum (UK), image trent_1000 used by written permission)

There are two or even three different shafts, one inside the other, for the low and high pressure sections, which allows the huge fan to rotate more slowly than the jet engine in the center, so it doesn't fly apart. Sometimes, gears are used.

The turbofan engine gives much more push (thrust) at takeoff, and modern planes can leave the ground at a steep angle compared to those of the 1960s. Also, they are a lot quieter. For my first flight on such a plane, in the 1980s, the difference was really exciting. I was being pushed back into my seat like I was in a Corvette! I dangled a menu in my hand for a makeshift pendulum, and compared its angle to that of the vertical edge of the window. That was about 15°, so I was accelerating at about 1/4 G. I had never seen or felt anything like that. Modern planes can climb at 1500–2000 feet/min, unheard of except in military planes in the 1960s, and we take it for granted.

The Rolls-Royce Trent 900 that powers the Airbus A380, the world's largest plane (Fig. 14) weighs a little over 7 t, and makes 40 t of thrust, nearly six times its own weight. It's a lot bigger than a man. Its compression ratio is 39:1. That big fan is about 10 feet in diameter and rotates at 3000 rpm. The tips of its blades are moving faster than the speed of sound! You can see that there's an immense amount of engineering that goes into a real jet engine!

Supersonic Commercial Jet Planes

The first commercial Supersonic Transport or SST was Russian, the Tupolev TU-144, which first flew in 1968, two months before the more famous French Concorde. Unfortunately for Tupolev and Russian commercial aviation, it crashed

Fig. 15 The Concorde. The droopy nose gave the pilot a better view for takeoff and landing (Eduard Mamet, permission under Creative Commons Attribution-ShareAlike 3.0 Unported License)

Fig. 16 The Concorde's Machmeter (George Moromisato used by written permission)

at the 1973 Paris Airshow, and again in 1978. Ultimately, it made over 100 commercial flights, but the Concorde stole the show.

Concorde (Fig. 15) was a beautiful plane. A product of EADS, the European Aeronautic Defence and Space Company N.V., it started transatlantic service March 2, 1969. The Concorde could fly New York to London in 2 h, 53 min—and the reverse in 3–1/2 h—less than half the time of a 747 or Airbus. Of course it only carried 100 people, but they were special, like the Queen Mother, the Queen, Prince Philip, Margaret Thatcher, Pope John Paul II, Joan Collins, Phil Collins, Sir David Frost, Tony Blair … you get the idea.

VIP's liked to brag that they could attend an 8 a.m. meeting in London, make a 9 a.m. meeting in New York, and return in time for a nice dinner at Le Gavroche. Over a million people used the plane.

It needed four Rolls-Royce SNECMA Olympus 593 engines, each making 19 t of thrust, to reach twice the speed of sound. This is called Mach 2 (2170 km/h) after the Austrian physicist and philosopher Ernst Mach (Fig. 16). It flew at 60,000 feet, 11 miles up.

Dinner was served on Royal Doulton bone china, as well it might for tickets that cost on the order of $8000 a flight. This is why mostly rich and famous people did it.

Range for the Concorde was 6667 km (4143 miles), enough to get from London to Dallas. Unfortunately, US regulators would not allow it to land in Dallas, supposedly because of the noise it would make going faster than sound. Of course, fighter jets do this all the time. In any case, that lucrative route was denied it. Ultimately, economics were so bad that one of the planes was being kept for spare parts for the others. Even with those prices it was not a business success. On July 25, 2000, Air France flight 4590 crashed in a blaze on takeoff after hitting a piece of junk on the runway, killing 100 passengers and nine crew and ending its career as a commercial airliner, after 27 years of great service. Today, there is no SST for you to experience.

Other Supersonic Jets

The SR-71 Blackbird (Fig. 17) was even faster. Now there's another beautiful plane! At Mach 3.3, it moved at 3450 km/h and that meant (Fig. 18) that it could fly London to Los Angeles in 3 h and 47 min at an average speed of 2900 km/h,

Fig. 17 The Blackbird (Image used by written permission from Lockheed Martin Corporation)

Fig. 18 Blackbird flights (Image used by written permission from Lockheed Martin Corporation)

Fig. 19 Shock diamonds in the Blackbird afterburner exhaust (NASA public domain)

making the leading edges of the titanium wings glow. It burned 130 gallons of fuel a minute to power its two J-58 engines, each of which made 16 t of thrust, with afterburner. This engine was much hotter than that in any commercial plane (Fig. 18). If you were following along there, you'll notice that the Concorde had four engines with that kind of thrust, but Blackbird was smaller. The Blackbird weighed 86 t at takeoff.

The neat part was that it could reach an altitude of 85,000 feet, almost up to the ozone hole. At that time, no missile could catch it. It went into service in 1966 and flew until 1998, about the same interval as the Concorde. Figure 19 shows the shock diamonds in the Blackbird's exhaust.

Pulse Jets

This invention is another way of solving the problem of having an engine that is open at both ends but still makes a jet in only one direction. Figure 20 shows the idea. You might say it solves it by cheating, because it's only open at both ends for long enough to let in the air. Then, you squirt in the fuel and ignite it with a sparkplug. When the explosion happens, it slams shut the little metal venetian blinds at the front so the jet can only go out the back. Simple, right? Kind of like the ram pump.

Unfortunately, its first practical use was as the engine that powered the German V-1 "Buzz Bomb" that caused so much destruction in WWII, the ancestor of our own cruise missiles.

You can still buy kits for this engine and make a jet powered bicycle! (Fig. 21). Warning: these things get really hot! Figure 22 shows one that makes 30 pounds of thrust.

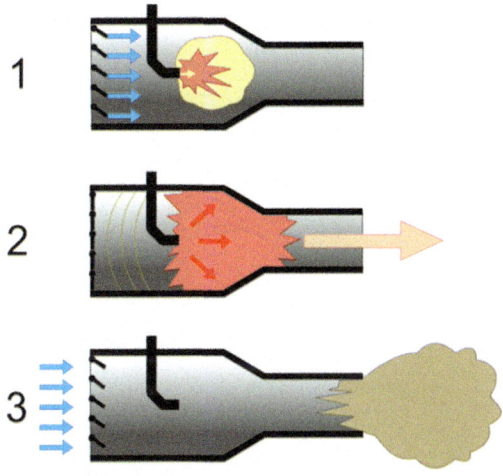

Fig. 20 Pulse Jet (Tosaka. Permission under GNU Free documentation license version 1.2)

Fig. 21 Jet powered bicycle kit (Robert Maddox, maddoxjets.com, by written permission)

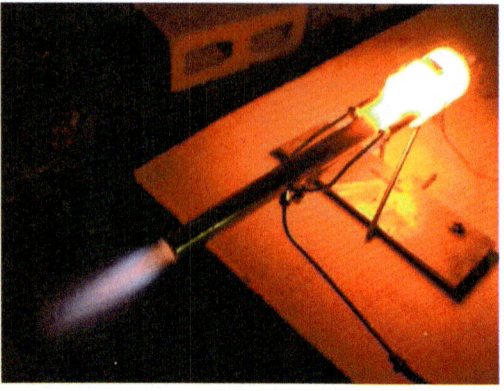

Fig. 22 Pulse Jets get hot! (Pulsejetengines.com/mediagallery)

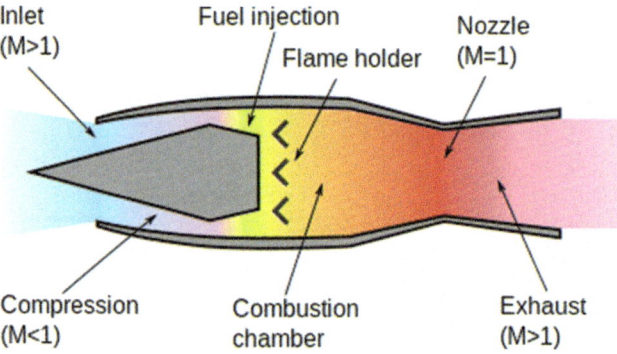

Fig. 23 Ramjet. M means Mach number (Permission granted under Creative Commons Attribution-ShareAlike 3.0, GNU Free Documentation License version 1.2)

Even More Advanced Jet Engines: The Ramjet

I haven't mentioned these before, because no flight you're likely to be on soon will use one. That's because ramjets have no thrust sitting on the ground, so you can't take off with them. They're only good for high speed flight, at their best at supersonic speeds, and like to work up to Mach 6. If you look at Fig. 23, you will think this IS finally perpetual motion. After all, they do call it a "flying stovepipe." There is no turbine or compressor! And that's the point: first you have to get this thing going very fast with an external rocket or something. Then, you will notice the same old principle as in the turbojet engine. The engine is already moving so fast that the air ramming into the narrow passage of the "inlet spike" makes the same high pressure that you got with the turbojet's compressor at the combustion chamber input. Then fuel is injected and burned and the hot gas expands and accelerates in a larger cross-section exhaust. Designing and operating these is very complex.

The Blackbird's J-58 engines that I mentioned earlier were actually "turbo-ramjets," using some features of both kinds of engines. Its inlet spike had to be moved front and back about three feet for it to work properly all the way from subsonic to Mach 3 flight. A turbo-ramjet might be the kind of engine on the wings if you do ride in an SST some day in the future!

Conclusions

When it comes to strength, power and speed, a properly designed machine can outdo an animal any time. Don't compete with a machine! Jet engines are an amazing invention. They make the Jet Plane possible. They are not rockets, because they get oxygen from the air to burn their fuel. There are many kinds of jet engine. People have made jet planes that went three times the speed of sound, so fast that the wings glowed, almost as fast as rockets! And ramjets can operate up to Mach 6. Wings lift everything from tiny mosquitoes to hummingbirds to eagles to helicopters to jet planes. They also make possible wind turbines and fans. The wings men have invented for commercial travel cannot flap up and down and are not under constant intelligent shape control so they have some disadvantages relative to living creatures' wings. But they do pretty well. Technology is wonderful, allowing us to do things tomorrow that we only dream of today.

Now let's see what is different about rockets!

Rockets

Like the jet engine, rockets are another invention that will make beings in the future impressed with our capabilities, if they're capable of being impressed. Rockets are exciting because they allow some of us to use them to explore space, not so much because ordinary people use them very much outside the fourth of July.

The first rockets were developed in the early thirteenth century by China. These were gunpowder devices, in design though perhaps not in size, just like the ones you buy for the fourth.

Rockets vs. Jets

The first essential difference between a rocket and a jet engine is that it carries its own oxidizer, instead of using oxygen from the air. This is a big plus for travel outside the atmosphere! The second is that, because air is only about 20 % oxygen while a rocket can offer pure oxygen to its fuel, the fuel can burn *very* hot, and that gives high speed and better efficiency. The third is that a rocket is functionally a simple device, not needing turbines to bring in and compress the air or to drive the compressor, so it can be quite compact and easily travel faster than sound.

Not to say rockets are simple - a Titan III-B Centaur is not a simple machine. Just simpler to understand how it works. Bring fuel and oxidizer together, light them, and you have a jet. In some kinds of rockets you don't even need a match.

This simplicity in basic function is why rockets were developed 700 years before jets, way before anyone thought of traveling outside the atmosphere. They were thought of as "fire arrows," and mostly were used to bring fire to enemy forts and facilities.

Today, there are two kinds of rockets: solid and liquid, illustrated in Figs. 1 and 2. Nozzles are used in both, but are usually relatively primitive in solid rockets. In liquid rockets, the nozzle has a specific shape indicated in Fig. 2, which is designed to convert the very hot gases in the combustion chamber into an exhaust in which as much of that heat energy as possible has been converted into high speed

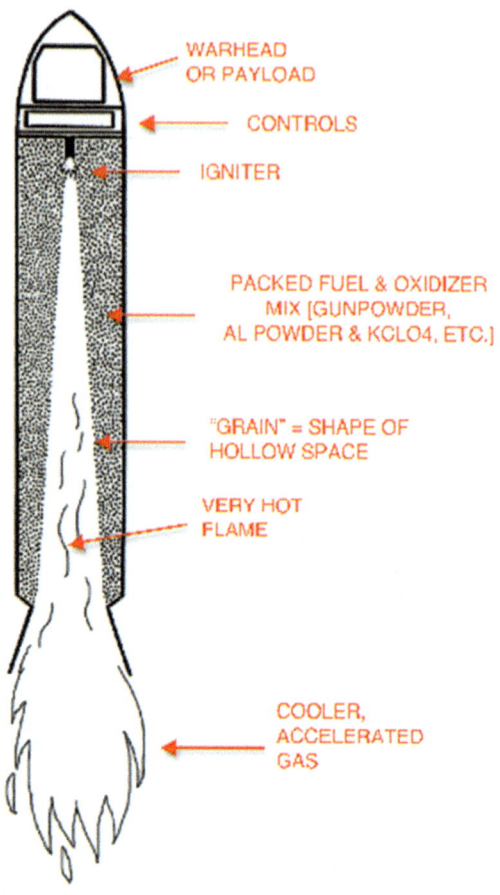

Fig. 1 Solid rocket (F. Wicke)

gas flow in one direction. Doing that pushes back on the rocket as efficiently as possible instead of wasting the energy in heat, light and gases expanding in all directions, like in an explosion.

Solid fuel rockets are not much changed in principle from Chinese skyrockets [*China*]. Even in this century, the solid rocket booster (SRB) that started the Shuttle on its voyage was just a giant skyrocket (the largest ever made), with a casting made of solid fuel and oxidizer inside that is lit off and just burns without any way to turn it off. In 1805, William Congreve discovered that, if you want to make the thrust constant, a since it's going to burn out into that shape anyhow, you'd better make the casting in the shape of a hollow cone to start with. This sounds like a terrible waste of space, but it worked well for a long time. Congreve rockets, weighing up to 30 kg, were responsible for the "rocket's red glare" mentioned in the US national anthem.

Today, with the Shuttle booster rocket for example, the grain shapes are more complex. Each of two of these 45-m-long rockets developed 14 million newtons of

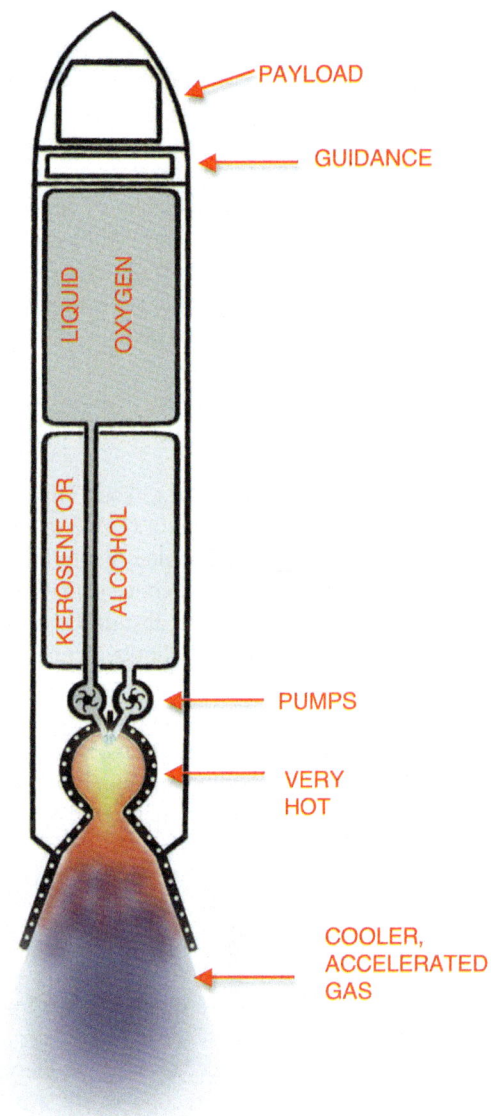

Fig. 2 Liquid fueled rocket (F. Wicke)

thrust, weighed 590,000 kg at launch and topped out at 67 km altitude after separation from the Shuttle. For this rocket, the fuel was aluminum powder (70 %) and ammonium perchlorate provided the oxygen (16 %). This released 4.9 million joules per kg of the total mixture when it burned, about like the 4.7 MJ/kg figure for gunpowder. These proportions just add up to 86 %; the rest was a touch of iron oxide catalyst and a plastic binder to hold it together.

Liquid fuel rockets have pumps to bring in the oxidizer and fuel. They have the advantage of being able to use the hottest chemical reactions known to man. Hydrogen and oxygen combine to form ordinary water, with the release of 13 MJ/kg from the right proportions of H_2 and O_2. You can see that's almost three times hotter than gunpowder. The hottest known liquid fuel reaction is hydrogen and fluorine, which gives 15 MJ/kg of the mixture. The only problem with that reaction is, it makes hydrofluoric acid, which is so violent that it's used to etch glass.

The American scientist Robert Goddard did the first experiments with liquid fuel rockets in 1926. He also figured out that you can cool that hot exhaust nozzle with the very fuel and oxidizer that are going into it. Other pioneers were Hermann Oberth and Konstantin Tsiolkovsky.

Reaction

By the way: experts at the New York Times in 1920 ridiculed Robert Goddard's work saying this: "That Professor Goddard, with his 'chair' in Clark College and the countenancing of the Smithsonian Institution, does not know the relation of action and reaction, and of the need to have something better than a vacuum against which to react—to say that would be absurd. Of course he only seems to lack the knowledge ladled out daily in high schools." You can also be pardoned for being confused about that 90 years later. But Goddard *was* right. Here's how it works: imagine I'm sitting inside a very large rocket out in space that is full of bowling balls [and that I'm wearing a space suit!]. I pick up a ball and push it out the back end of the rocket as hard as I can. In a bowling alley on Earth, my legs would brace against the floor. In the rocket, my back had better be up against the wall or I will go flying backwards, right? Sooner or later, I'm pushing on the rocket, making it go forward faster. If instead of bowling balls the rocket is squirting out hot gases, the same thing happens.

That's all there is to it. OK, here's a more practical example. Have you used a power washer or fired a shotgun? When you squeeze the trigger, what goes out pushes back on you. The water or shot is not pushing against the air. It would work as well in vacuum.

The first truly modern rocket was the German V2, a terrifying ballistic missile used to great effect in the second world war (Fig. 3). Over 3000 were launched, primarily at London and Antwerp. The flight took about four minutes, reached 97 km altitude and a speed of 5400 km/h. The V2 had every one of the parts of a modern rocket, including an inertial guidance system [see *Lasers*]. A complete V2 including the guidance system can be seen today in the center of the spiral staircase in the Deutsches Museum in Munich. It used a significant part of the German potato crop in the final years of the war to make its alcohol fuel, plus liquid oxygen. It took 30 t of potatoes to fly one V2. That made the German people pretty hungry. It also took about 20,000 slave laborers working nonstop to build them in underground caves. Even the shape of it is iconic, having appeared on the covers of dozens of science fiction magazines and in movies of the 1950s as, for example, the George

Fig. 3 The V2 (F. Wicke)

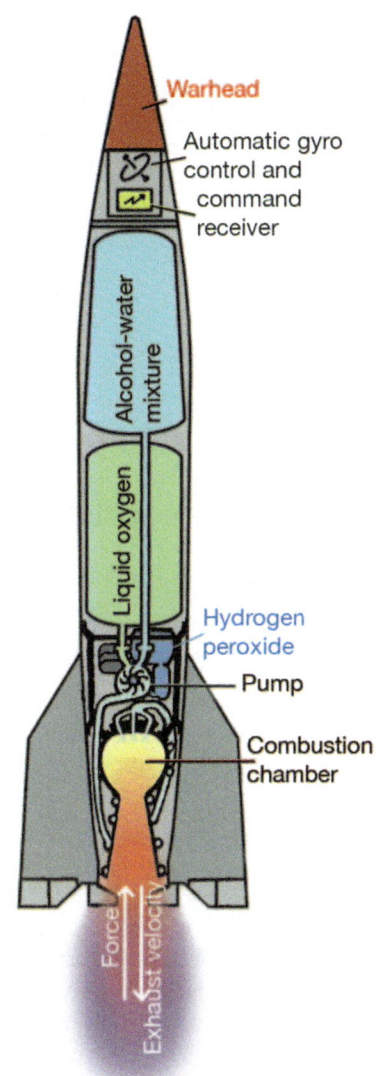

Pal film "Destination Moon". Many of them were moved to New Mexico for tests, but only about half were ever flown successfully. One that was took the first picture of the Earth from space in 1946 (Fig. 4).

After the war, the USA got primarily German *scientists* who developed the V2, like von Braun and others in *Operation Paperclip* [see *Jet plane*], but the Russian equivalent took primarily the *technicians* who actually knew how to make those things work, giving the Russians a great initial advantage. In 1959, Huntsville, AL (the home of the Army's Redstone Arsenal) was filled with names like Beichel,

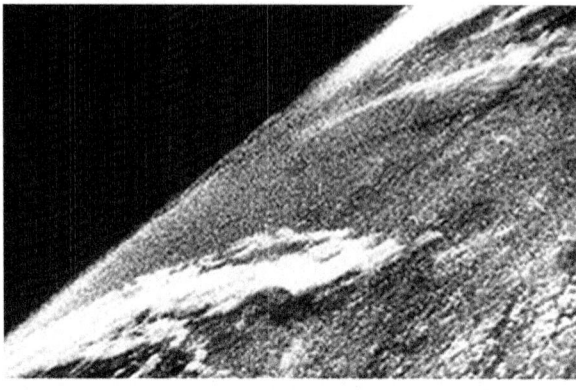

Fig. 4 First picture of Earth from space, taken from a V2 launched from New Mexico (White Sands Missile Range, US Govt., Public Domain)

Dahm, Debus, Hohmann, Maus, Neubert, von Sauma, Stuhlinger, Zeller, ..., and von Braun. It is said that one heard German more often than English in Huntsville in those days. This choice had consequences all the way forward to the launch of Sputnik.

As late as 1958, when President Eisenhower founded NASA from the older NACA, a joke I remember went "How are the rockets at Cape Canaveral like NASA workers?" "They won't work and you can't fire 'em!'" Of course, neither of those is true today, but you can see movies about the early failures on the internet.

Modern rockets mostly use either kerosene and oxygen or hydrogen and oxygen. Kerosene is a lot easier to handle than liquid hydrogen, which must be kept at its boiling temperature of 20°K [see *metric system*].

Satellites

Now, let's look at what rockets can do!

The very first manmade Earth satellite, Sputnik, was launched by the Russians in October, 1957 (Fig. 5). Its mass was 84 kg. You want to know how old I am? I was a freshman at MIT when that launch happened. In Physics class, some people said it was a bluff, probably just a balloon. Others said you couldn't put a balloon in orbit, and we argued. They were not correct. And it wasn't! It weighed 84 kg and came back to Earth and burned up in a couple of months because it wasn't high enough to avoid the atmosphere for very long.

Sputnik

More shocking yet to the average US person, accustomed to us being first in everything, the Russians followed that with Sputnik II 1 month later, carrying the first living being (ill-fated Laika) into orbit. At 508 kg and 2×4 m, this one was huge by comparison to anything any other nation put up until the late 1960s. Even Telstar, launched in 1962, was only 77 kg. Not only that, the entire assembly that went into

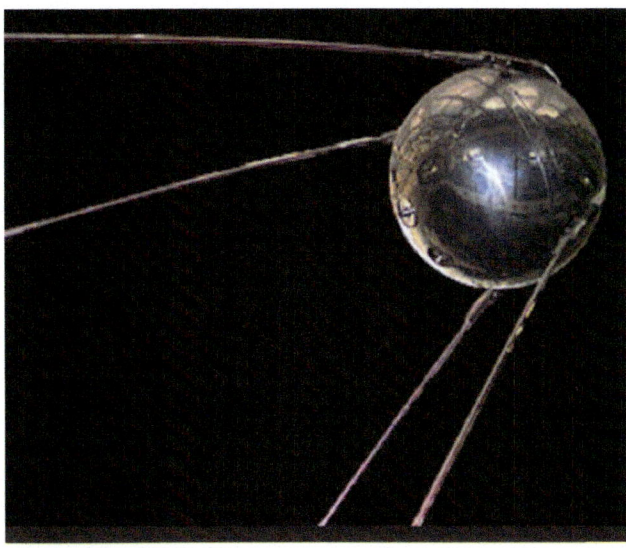

Fig. 5 Sputnik (National Air and Space Museum Public Domain)

orbit with Sputnik had a mass of about 8000 kg. That's eight tons! They didn't intend that—Sputnik II failed to separate from its Blok-A final stage, and this had an empty ("dry") mass of 7500 kg. The bundle of engines that drove the first stage generated 7MN of thrust (enough to launch a 700 t liftoff mass) (Fig. 6).

Sputnik II stayed in orbit for half a year. It was intended that Laika would be poisoned by her last meal before asphyxiating when the oxygen ran out. Instead, she died more quickly from heat and stress.

Oh, but wait! The show was not over. On May 15, 1958, Sputnik III was launched, with 1300 kg of useful payload, almost three times the 508 kg of Sputnik II. And it stayed in orbit for 2 years. This had all happened within 7 months!

The first US satellite was Explorer I, launched just 3 months later on January 31, 1958. At 14 kg, it was smaller, but its orbit was higher, so it continued to send out beeps for 12 more years before re-entering the atmosphere. It was actually quite useful, and discovered the Van Allen radiation belts around Earth.

In March, we launched another one called Vanguard, only 1.5 kg, which was derisively called "the Grapefruit." Because its lowest point is 650 km, way above the atmosphere's friction, and because it is the first solar powered satellite, it is still in orbit now, 55 years later, happily beeping away, and will be for the next 200 years!

In those days of the Cold War, everyone also realized these rockets could deliver hydrogen bombs to the USA. The near-hysteria that resulted was the direct reason for the creation of NASA. It was also a stimulus for the complete, frantic overhaul of elementary school mathematics education in the USA, which is why you have to learn set theory today!

Fig. 6 The Russian R7 rocket which launched the Sputnik series (Heriberto Arribas Abato Creative Commons License, Wikimedia Commons)

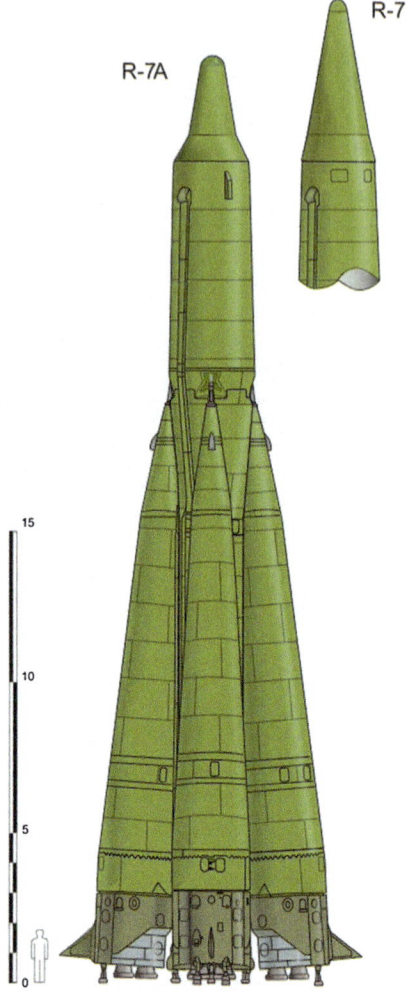

The US program did not develop an engine as big as the Russian ones until the Saturn V rocket during the Apollo program, whose 5 Rocketdyne F-1 engines together delivered 34 MN thrust at liftoff (Fig. 7).

These mighty engines (the most powerful in history) were finally ready to fly in 1965, 7 years after Sputnik III. They burned kerosene and liquid oxygen. The main person responsible for the success of that program was Wernher von Braun (Fig. 8), the man who developed the V2 in Germany, and who finally realized his dream of sending a man to the Moon in the Apollo mission of 1969. It was also a success for *Operation Paperclip*!

And, in one of the most ironic twists of history, the F-1 engine and the Saturn rocket are no longer built. A friend of mine who worked on the team that built it assured me personally that the expertise for doing that is lost and that it would take

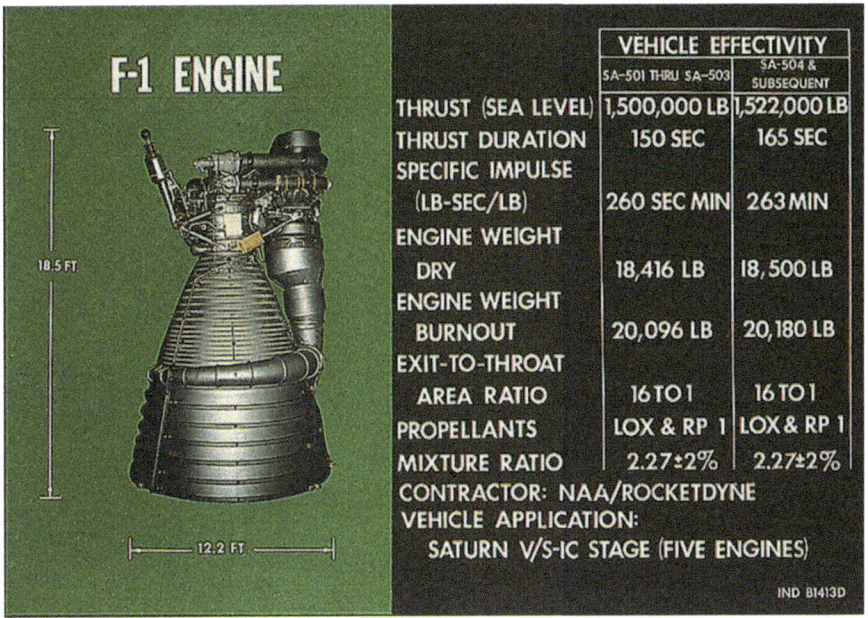

Fig. 7 Rocketdyne F-1 (NASA Public Domain)

Fig. 8 Wernher von Braun (NASA Public Domain)

a long time to regain. We have also stopped building Shuttles. Meanwhile, a variant of the Russian engine that launched the Sputniks now carries our astronauts to the Space Station with the Soyuz crew module! Reminds me of the Aesop fable about the tortoise and the hare…

From the Moon to Mars and Far Beyond

Many of us can remember and the rest of us can read about the first landing on the Moon in 1969. "A small step for a man…," but that's as far as manned exploration has gone. There's a good argument on both sides of the question about whether manned or robot exploration is best, and we won't get into it.

What is a fact is that such huge boosters are not required to explore Mars, or even to launch items like the Voyager spacecraft, which has now moved on out of the solar system. Launched in 1977, the 722 kg Voyager is 18 billion km from its starting point here, 125 times the distance of Earth from the sun. Radio signals take 16 h to get to us from it at light speed! The reason it can still work out there where the sun is a distant star and it's very cold and dark is its 160 W plutonium oxide power supply.

Even this winds down. Next year, it will turn off its tape recorder, the year after that its gyros and, in 2020, everything. Whatever it finds after that, it will no longer be able to tell us.

It sent us our first close-up pictures of Jupiter more than 30 years ago (Fig. 9), and lots of data since then. It is definitely something that will survive life on this planet!

Fig. 9 Jupiter as seen by Voyager (NASA Public Domain)

A magnificent animation showing the utter complexity of modern launches to Mars is here.[1] In that animation, you will see the "sky crane" lift away and dutifully crash itself on a distant hillside—something a manned alternative would not do.

We have explored other planets, too, and even their moons. You can be fascinated for days looking into this yourself if you have a deeper interest!

Exotic Rockets

If you want to go very far with a spacecraft, and get results while people on Earth still care about them, you had better go very fast [*Weird Reality*].

But you can only carry so many bowling balls. So: it's clear you had better push each one out as fast as you can. If you don't, they'll all be gone before you gain much speed.

Ion Engines

One way to do that is to electrify a gas like xenon until it is a cloud of electrons and ions, and then push the ions out into space with an electric field. Of course, you must later add some electrons so the cloud has zero net charge, or the ions will just come back at you. The result is the engine "Deep Space 1" which took a NASA spacecraft out to a rendezvous with Comet Borrelly in 2001. Its exhaust moves at about 35 km/s, faster than the Earth goes around the sun. Figure 10 shows what an ion engine exhaust looks like.

Fig. 10 An ion engine at work. NASA Public Domain

---

[1] http://www.engadget.com/2011/04/13/nasa-animation-depicts-curiositys-soft-landing-on-mars-courtes/.

Laser Space Propulsion

You can also shine a very, very bright laser on those (tiny) bowling balls, turning them into a very hot gas. You can get even higher speeds than with ion engines that way. My partners and I have made an engine that uses this idea (Fig. 11). You can also use lasers to propel something else far away from you. If you use a laser to make a jet on a piece of space debris circling the Earth, you can slow it just enough that it will drop into our atmosphere and burn up.

Laser "Lightcraft" that fly in the air also exist and have reached a height of 70 m in the New Mexico desert. Laser space debris removal depends on the same thing.

Photon Rockets

The very highest exhaust speed physics permits is the speed of light. You'll never feel it, but if you shine a flashlight there is a very tiny push back on you from launching that beam! Believe it or not, this idea has been proposed for space travel. However, because it requires many long-distance bounces in order to work it is not yet clear it will operate over long distance. You can measure this force in the lab.

Thermonuclear Rockets

Starting with an idea in the 1940s and worked on intensively in the 1950s, the original "ORION" program intended to get exhaust speeds in the 35 km/s range with a lot more thrust than ion engines will ever make by setting off an atom bomb every second behind the spacecraft. The ship was supposed to weigh 4000 t and reach Mars in 30 days! Although there is little doubt that it would have worked (at least for a few seconds!), people on Earth might have been upset by the idea of

Fig. 11 Laser Plasma Thruster operating in a vacuum chamber (C. Phipps, Photonic Associates, LLC by permission)

launching such a thing. There are other problems beside fallout (like knocking out communications) that occur with a nuclear explosion in space anywhere near the Earth.

Satellites: What Holds Them Up?

Back to satellites for a moment: what holds them up, and why is there zero gravity on the Space Station?

If you always wondered what holds satellites up and why gravity disappears onboard the Space Station, your answers are here!

It's not complicated. When you tie a rock on a string and whirl it around your head, it pulls away with a force that increases in proportion to the square of the rock's speed, and to the mass of the rock divided by the length of the string. It's called a centri*fugal* force, because things experiencing it try to get away from the center. A practical application of this is that, if you're zooming onto a freeway exit at 80 mph, it's four times as hard for your tires to keep you from flying off the road as it would be at 40.

Imagine that the rock is a satellite and that gravity from the Earth is the string. Imagine that the string is tied at the center of the Earth. Of course, it would be burned up quickly, because it's hotter than hell down there! Worse, it turns out it's a solid iron ball at the center. But just imagine. Gravity drops off as you move away from the Earth, like the length of that string squared. The force of gravity drops off with distance from the center faster than centrifugal force, *so they balance at some distance*. Of course, if the satellite speed is too slow, that distance might be under the surface of the Earth, and then, Houston, you've got a problem! So centrifugal force holds a satellite up. When centrifugal force and gravity are balanced, you have an orbit — at least until you run into something, and you feel zero gravity onboard.

Those of you who are lucky enough to have had algebra can figure out that this critical speed squared for a circular orbit is proportional to a constant involving the strength of gravity, divided by the distance to the center of the Earth. That means the speed goes down by the square root of the distance. But the distance around the circle of the orbit is proportional to the distance, so the time a satellite takes to go around the Earth (called the period) increases with the 3/2 power of the distance from the center. See *Modern Science* for more details.

Zero Gravity Environments and Free Fall

Now, back to the second question: how and why can you make gravity disappear? You cannot turn gravity off. So far as we know, you can only experience zero-G by being in free fall, in orbit, or by being far away from any planets or stars.

Of course, you don't have to be in a nice, circular orbit to experience zero gravity. You can be jumping off a building (briefly) or dropping like a rock toward Earth. People actually do that, in an Airbus that flies in a parabolic arc up and then down,

Fig. 12 Zero gravity (Jurvetson Creative Commons 2.0 Generic Attribution-ShareAlike License)

like a slug shot from a cannon. A really experienced pilot pulls them out of the dive just in time. This is a lot cheaper than a space launch, and gives the passengers 25 s of zero-g to see what it feels like to be a real astronaut. Doesn't sound like much, but count out 25 s and you'll see how long it is (Fig. 12). After that—you'd better buckle up because the 1.8 G pullout from that dive will be tough!

On board the Space Station, gravity is not absolutely zero everywhere, but only at one spot. The scientists there call it a "micro-G environment." If you move toward the Earth from the zero-gravity spot by just ten meters, you would feel 0.3 micro-G's because you're still moving at the same speed as the Station, but you're closer to the center of the Earth. If you move the other way, you'd be 0.3 micro-G's lighter! You won't feel the difference.

You can do lots of interesting things in a micro-G environment. Figure 13 shows one—a huge glob of water floating in front of astronaut Story Musgrave and refracting the image of his face.

The fact that long spindly things pointing toward the Earth in orbit will be "heavier" at one end than the other because of micro-G's keeps them pointing toward Earth is called "gravity gradient stabilization."

Geosynchronous orbit is 35,882 km altitude above the Earth's surface, where the satellites are located that send you your TV programs. Up there, the speed where gravity disappears is only 3.07 km/s. It's slower because you're farther from the Earth and gravity's pull is less. It's so slow that the time it takes to go around the Earth is 1 day. So, as the Earth turns, these satellites stay directly overhead! This is a useful location if you want to stay exactly aligned with everyone's satellite dishes.

Closer in, it's quite different. For the Space Station, if they didn't keep reboosting it, the Station would re-enter and burn up in a year or so. It uses 7000 kg of fuel per year for reboost, to avoid debris, and to keep it pointed in the right direction.

Fig. 13 Water in microgravity (NASA Public Domain)

A very strange thing happens if you want to go from the orbit of the Space Station for example, at 370 km above the Earth to geosync. To get there, I turn on my rocket motor and use a lot of fuel, but I end up going slower!

Going farther out, the moon is in its own zero-g environment at a distance of 384,000 km, where it moves so slowly that it takes about 28 days to go around the Earth at 1 km/s.

You never feel the gravity of the sun, right? The whole Earth is a Space Station circling the sun in an orbit with a 150 million km radius, but because it's circling at 29.8 km/s, we live in a zero-g environment with respect to the sun. Or the center of the Earth does. If you're closer to the sun or farther from the sun, you feel a slightly different force. The oceans are 12,800 km closer to the sun on one side than the other, feel more or less pull from the sun and that helps to make tides!. It takes us a year to go around the sun at that speed. You knew that, I hope! Right now, you're going at this tremendous speed, almost four times the speed of a low-orbit Earth satellite, and you don't even feel it. The speed needs to be that large because we're orbiting the Sun, not the Earth, and it has a tremendous mass.

Actually, if you were hanging on a string in space, not moving, far away from the Earth, you would only weigh 60 g (about 2 ounces) in the sun's gravity. That tiny force, combined with the pull of the moon, makes the tides on Earth, and also keeps the Earth from flying off into the darkness of outer space.

Zooming Along Without Feeling a Thing

There's more to the story - you're going even faster than that. Let's add it up: if you live on the Equator, you're going 0.5 km/s toward the East because the Earth is rotating, almost 30 km/s in the yearly trip around the sun and—get this—you, the Earth, the other planets and our Sun, are rotating with the whole Milky Way Galaxy

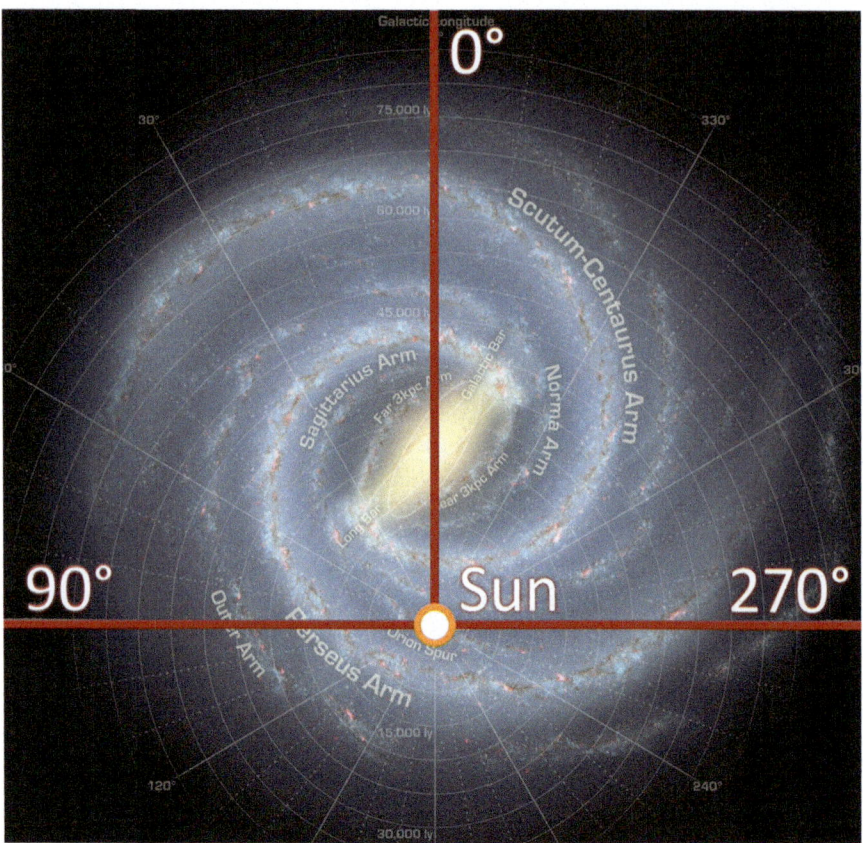

Fig. 14 Our Galaxy. A *Black Hole* is in the center (NASA Public Domain)

at about 215 km/s (Fig. 14), so that the "galactic year" is about 250 million years. Then, the local group of galaxies is itself moving relative to the microwave background [see chapter on *Cosmology*] at 627 km/s. Don't get seasick when you go to sleep!

We now know that we are far from being the center of anything.

Space Junk: The Downside of Satellite Launches

Since the dawn of the space age, a lot of low-Earth-orbit satellites have been launched, and, when the fuel runs out that keeps them pointing in the right direction, or the batteries die, etc., you have a piece of space junk instead of a useful satellite. Not only that, but the last stages of rockets that brought them up there, explosive bolts and so on are also in orbit floating around. The rockets sometimes have fuel

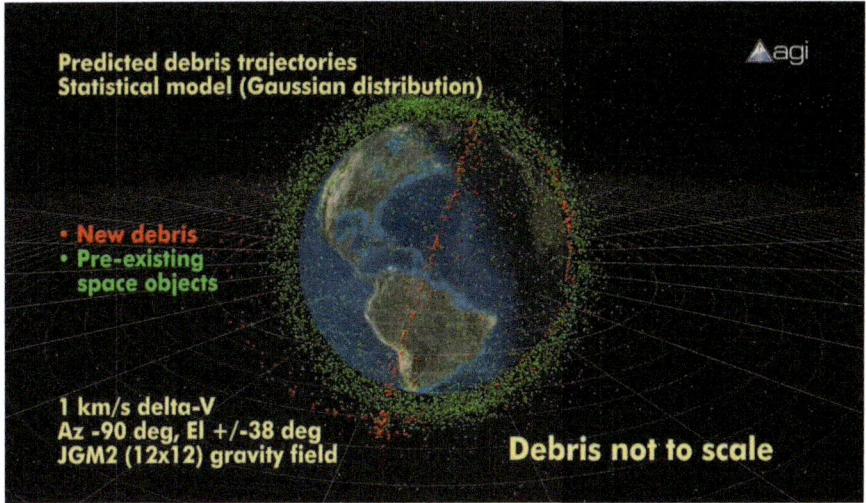

Fig. 15 The Kosmos–Iridium collision. New debris created by the collision are in *red*. The collision was unexpected (Image courtesy of Analytical Graphics Incorporated, by written permission (www.agi.com))

left over, just in case, and explode years later. Sometimes they hit each other and explode, or are deliberately blown up.

When a Russian Kosmos satellite collided with a US Iridium satellite in 2009, thousands of new pieces of space junk were created (Fig. 15). This is happening more and more often. There are now about 19,000 large pieces of debris (up to several tons in mass), and about 250,000 1–10-cm pieces in low Earth orbit. Near misses from these force our astronauts to hide out in the smaller Soyuz module [when it is available] from time to time in order to have a better chance of surviving if a piece has too large a chance of hitting the Station. Space has to be cleaned up. Laser space debris removal systems can be used to do that by making the debris slow down just enough to burn up in the Earth's atmosphere [see chapter on *Lasers*].

What One Invention Will Survive Us?

Geosync satellites are far enough from Earth that there is almost no air resistance left to slow them down. If civilization died out or destroyed itself down here, those satellites would continue to circle for 100 million years! They likely won't work that long, but they will be one monument we leave to future life forms (Fig. 16). I already mentioned *Voyager*.

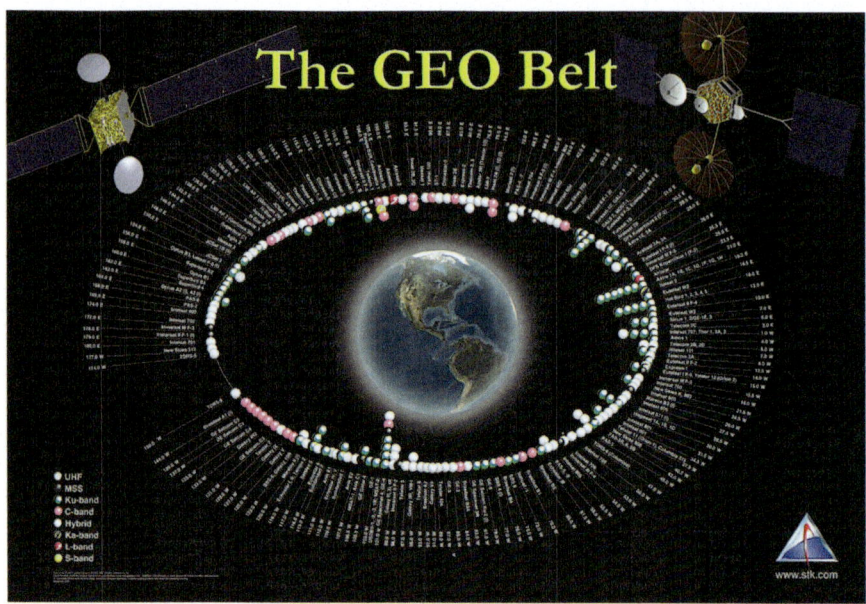

Fig. 16 "Ringworld" in GEO (Image courtesy of Analytical Graphics, Inc. by written permission)

Conclusion

It is truly magic for apes on this planet to have evolved to the point where they dreamed of making their own satellites, learned how to make the earthshaking machines required to put them up, and even how to get on board our own Moon and look back at Earth. In this book, we can only talk about a few of the clever things that were done along the way. I hope some of it strikes a spark and leads you to explore further!

Next: the fascinating topic of electromagnetic waves!

Electromagnetic Waves

Have you heard someone talk about EM waves? Have you wondered what they are? Why was it 150 years ago before anyone knew they existed?

Table 1 shows *21 powers of 10*—in wavelength, frequency and photon energy, as we go from cosmic rays to TLF radio. And yet electromagnetic waves are *all the same*. Light, cosmic rays, X-rays, cellphone radiation: the only difference is the wavelength! EM waves go spinning through space at the speed of light and are made up of a vibrating electric field that's (almost) always perpendicular to a vibrating magnetic field (there are always weird exceptions that you wouldn't want to know about), and spinning to the right (Fig. 1) or the left (Fig. 2), or a combination of these right and left waves into linearly or randomly polarized light.

A combination of equal amounts of right and left spinning waves is a linearly polarized wave (Fig. 3). They all obey the same laws. Depending on their energy or wavelength, they have vastly different effects when they go through materials. That's a lot for two pages! Let's go back.

Electric field? *Magnetic field*? "What's a field—Waves of *what*?" you well might ask. That was a big question up until 1887 and even later. If it's a wave, something has to be moving, right? Not really. Nothing moves that you can see or feel. You know light goes through the vacuum of space, from the Sun to us as an example, so it can't be *stuff* moving, right?

EM waves are not waves of any material, like ocean waves are. Instead, they are waves of the vacuum, of space. These kind of waves are both electric and magnetic waves (Figs. 1, 2, and 3), so they're called *electromagnetic*. You might say the vacuum electrifies and magnetizes a bit when one goes by. What these waves are is hard to understand, even today. But what sets a limit on the speed of light is not hard to understand. The limit is set by how much effort it takes to electrify and magnetize vacuum. It's just like how the tone a guitar string makes when you pluck it is set by how much effort it takes to pluck the string, and that depends on how tight you stretch it. You've done that. It's just the plucking that's hard to grasp, so to speak!

It's only been about three times the age of your grandpa since people even knew there were EM waves. In 1865, a really bright physicist named Jim Maxwell figured

Table 1 EM waves and their characteristics (see *Numbers, Lasers, Metric System*)

Name	Main use	Abbrev.	Wavelength	Frequency	Energy of one photon
Cosmic rays	–	–	<100 fm	>3Z Hz	>12 MeV
Gamma rays	–	γ	0.1–10 pm	0.03–3 ZHz	0.12–12 MeV
Hard X-rays	Laser fusion	HX	10–100 pm	3–30 EHz	1.2–120 keV
Soft X-rays	Making IC's	SX	0.1–10 nm	0.03–3EHz	0.1–1.2 keV
Vacuum ultraviolet	Chemistry	VUV	10–200 nm	1.5–30 PHz	6.2–124 eV
Shortwave ultraviolet	Lasers	UVC	200–260 nm	1.2–1.5 PHz	4.8–6.2 eV
Middle ultraviolet	Chemistry	UVB	260–315	0.95–1.2 PHz	3.9–4.8 eV
Longwave ultraviolet	Growing things	UVA	315–400 nm	750–950 THz	3.1–3.9 eV
Visible	Seeing	VIS	400–720 nm	420–750 THz	1.7–3.1 eV
Near infrared	Gas lasers	NIR	0.72–1.5 μm	200–420 THz	0.8–1.7 eV
Middle infrared	Heating	MIR	1.5–5 μm	60–200 THz	0.25–0.8 eV
Far infrared	Lasers	FIR	0.01–0.1 mm	3–30 THz	12–120 meV
Terahertz	Inspecting you!	THz	0.1–1 mm	0.3–3 THz	1–12 meV
Extreme high frequency	Sat TV, Astro.	EHF	1–10 mm	30–300 GHz	0.12–12 meV
Super high frequency	Sat TV, phones	SHF	1–10 cm	3–30 GHz	12–120 μeV
Ultra high frequency	TV, phones	UHF	0.1–1 m	0.3–3 GHz	1.2–12 μeV
Very high frequency	TV, FM radio	VHF	1–10 m	30–300 MHz	0.12–12 μeV
High frequency	Shortwave radio	HF	10–100 m	3–30 MHz	12–120 neV
Medium frequency	AM radio	MF	0.1–1 km	0.3–3 MHz	1.2–12 neV
Low frequency	Communication	LF	1–10 km	30–300 kHz	0.12–1.2 neV
Very low frequency	Navigation	VLF	10–100 km	3–30 kHz	12–120 peV
Super low frequency	–	SLF	0.1–10 Mm	30–3000 Hz	0.12–12 peV
Extreme low frequency	Early Navigation	ELF	10–100 Mm	3–30Hz	12–120 feV
Tremendously low freq. (!)	–	TLF	>100 Mm	<3 Hz	<12 feV

Note: (Last column) The energy units are electron-volts (eV). 1 eV is 0.16 attojoules (aJ). (Top row) Another way to state the energy of a 12 MeV cosmic ray is 1.9 pJ. Sounds small, but it's enough to make those funny bubble tracks in the plastic of your airplane window. And in you! For particles, 12 MeV is a lot of energy! It's worse on the Space Station, where astronauts sometimes see flashes in their brains. SLF waves are the size of the Earth. In the mid 1960s, before anyone could think about GPS satellites, the Navy wanted to let ships and subs use ELF to locate themselves, and created huge transmitter stations for the purpose. Another use for SHF is microwave ovens and radar. See *Lasers* for details of the visible spectrum

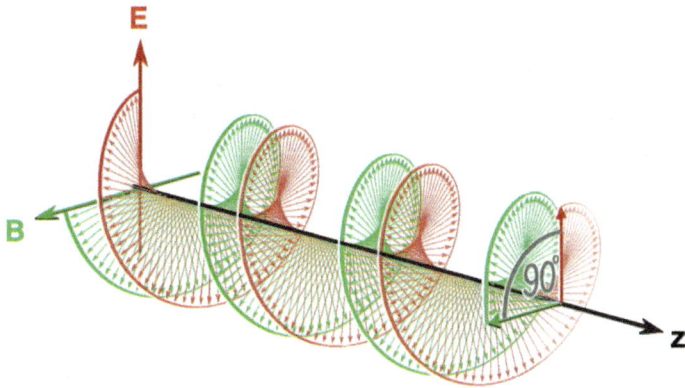

Fig. 1 Electric field vector (red) of a right (RCP) circularly polarized electromagnetic wave [magnetic field vector (green) is perpendicular and clockwise viewed from source] (C. Phipps)

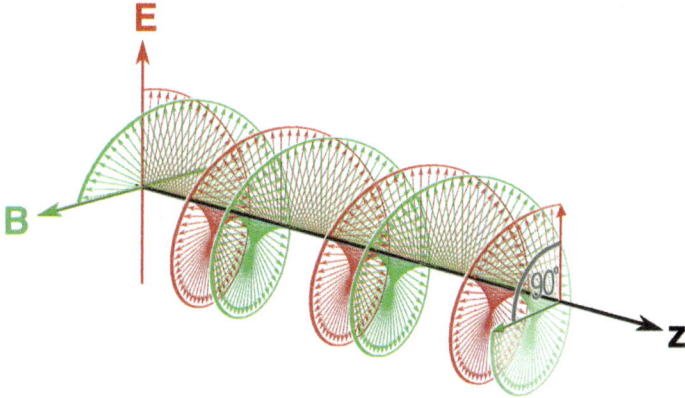

Fig. 2 Electric field vector (red) of a left (LCP) circularly polarized electromagnetic wave [magnetic field vector (green) is perpendicular and clockwise viewed from source] (C. Phipps)

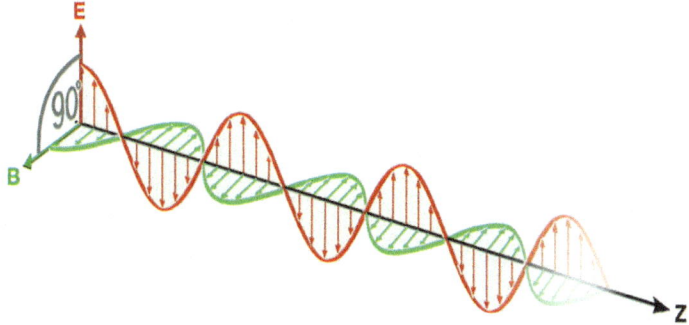

Fig. 3 Linearly polarized electromagnetic wave, a combination of equal amounts of RCP and LCP (C. Phipps/F. Wicke)

out with mathematics as well as experiments that radio and light are the same kind of waves with different length, going at the same speed.

Back in 1887, there were no satellites or astronauts, so they couldn't tell what speed radio or light waves used in space. Scientists invented invisible stuff they called "luminiferous ether" (not the kind that knocks you out) that the waves moved through out there and all around us here. This stuff was supposed to be fixed in the universe. Of course, nobody ever saw it or felt it—another mythical creature!

Michelson–Morley Experiment

That year, two scientists called Michelson and Morley imagined an amazing experiment to see whether this stuff existed. They used the Earth's own speed around the sun [it goes 30 km/s on its orbit around the Sun!], when it's going in opposite directions 6 months apart to test that idea (Fig. 4).

They used an *interferometer* which is even today called a Michelson interferometer (Fig. 5). When the light gets back from its travels along path 1 and path 2, it will interfere with itself at the detector. Let's say you set the path lengths for a "null," where the two beams cancel each other out. Now, point path 1 along the direction the Earth moves in the Spring and again in the Fall. If there were some material they figured they'd get a fraction of a wave difference between the two seasons because of the reversal of the "ether wind." They were thinking about it as if light were a sound wave going through some material. How would they tell which is the right direction? They didn't worry about that—they just put the whole thing on a pool of mercury so they could turn it slowly and see if there was a difference because of the

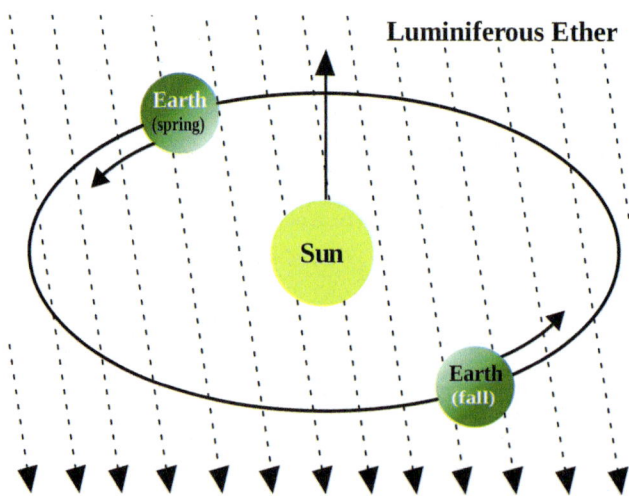

Fig. 4 Michelson–Morley experiment. Supposed "ether" is shown with *dotted lines*, and you can see how the Earth and the Michelson lab change their velocity through it from Spring to Fall (Cronholm144 GNU free documentation, Creative Commons Attribution-ShareAlike license)

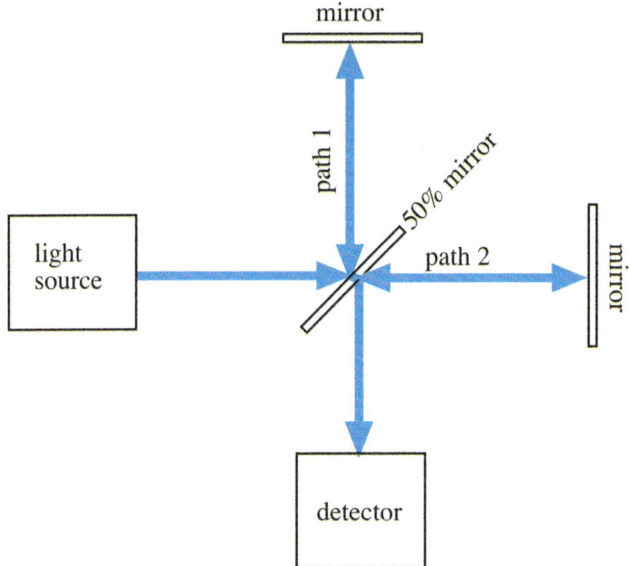

Fig. 5 Michelson interferometer (C. Phipps)

way path 1 and path 2 pointed, in June and again in December. That way they didn't have to keep the interferometer stable for 6 months. They predicted 0.44 waves of difference. I hope they didn't breathe deeply around the mercury pool.

They found nothing.

But! *In the process, they set the wheels in motion for the whole field of relativity*!

It's the most famous failure of modern physics! Paths 1 and 2 were 11 m long, and they had to take account of another factor (Fig. 4). This is a little hairy, but just think about it. If this setup can measure what it claims to, then in the time it takes light to travel down path 1 and back to the beamsplitter, the whole setup has moved! The beamsplitter will be in a different place and if you're going to interfere those beams, you have to tilt the splitter a bit and move the detector down the line a bit so all the beams end up at the right place. In the Figure, the angle θ is *really* exaggerated so you can see what's going on. Actually, it's only 0.006° when v is 30 km/s and c, the speed of light, is 300 Mm/s.

Five years later (1892), the scientists Lorentz and Fitzgerald (and some others) threw water on the whole idea of the Michelson–Morley experiment.

Here we need to think of ourselves standing in space as Earth and Fig. 4 experiment zoom by. Lorentz and Fitzgerald *guessed* that things shorten along the direction of motion when they're going by us very fast and, as we said in *Modern Science*, time changes too. The setup shrinks enough that θ disappears! In that case, poor Michelson and Morley wouldn't have measured anything, even if there were an ether. What a bummer, after all that work!

And Lorentz/Fitzgerald were correct. But that's science! Failure often produces great knowledge. Einstein was going home on the train one day from his job at the Swiss patent office when his train zoomed past another one going the opposite direction, and he thought about that angle θ. This all led directly to his theory about space and time, starting with Maxwell's equations. By 1905, we didn't need an ether, because *Relativity* explained everything.

Vectors and Fields

But I'm getting ahead of myself. Fields? Let's start with vectors. See that "**v**" in Fig. 6? It's a vector, which just means that it has a size v and it points somewhere, in this case to show which way something is moving *and* how fast. An electric field **E** or magnetic field **B** is the same sort of animal as **v**—unlike T for temperature or p for pressure, which just have a size but no direction.

Maxwell's equations? Well, actually, Heaviside's equations in the form we use today. But that's another story, maybe even another book, about how a guy without enough degrees after his name gets ignored. All we know Oliver Heavisde for today is a layer in the ionosphere.

What is a field? Just a bunch of vectors. Figure 7a shows an electric field **E** between two metal plates connected to a battery. If there were a proton between the plates, it would be pushed along just in the direction pointed to by **E**. Figure 7b shows the magnetic field **B** between two magnet poles. Magnetic push **F** is weird. Instead of being along **B** field, the push on the wire between the magnet poles is

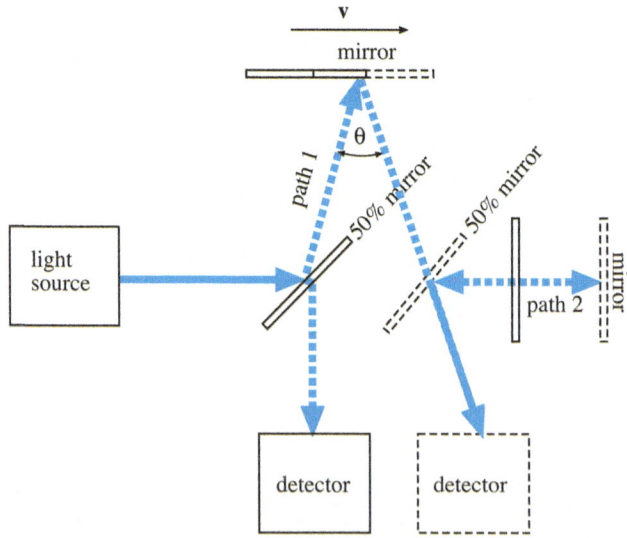

Fig. 6 Moving Michelson interferometer (C. Phipps)

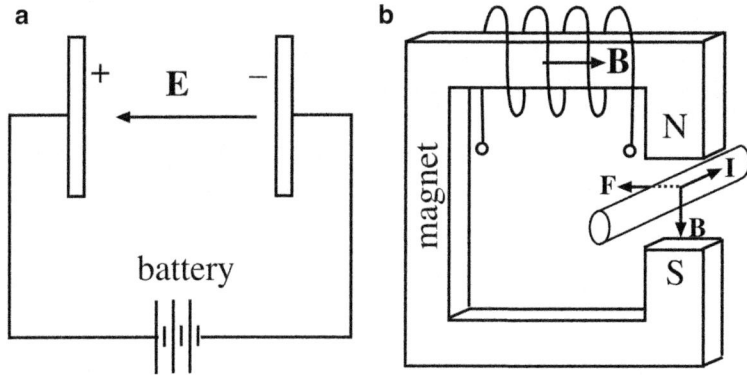

Fig. 7 Electric and magnetic fields. The magnet could be permanent, or it could be an electromagnet, with a current-carrying coil wrapped around it. "N" and "S" are for north and south poles (C. Phipps)

perpendicular to both the current **I** and the field **B**. This is the whole basis of mechanical meters, electromagnets and motors [see *Electricity*]. Why is it called **B** instead of **M**? Just an accident of history. Not German or Russian history as happens sometimes, but English history—Maxwell himself named it that.

Maxwell's Equations

Now then. Despite the heading of this section, I will not drag you through the equations, only explain them here. There are just two really important laws involved.

The first says (Fig. 8) the "circulation" of **B** is proportional to the current. If it's an AC current coming from a bunch of electrons surging back and forth because of a rapidly changing **E** field, then *the circulation of **B** is proportional to the rate **E** changes*, and goes back and forth with it.

The second says *the "circulation" of **E** is proportional to the rate **B** changes.*

These laws have a natural beauty because they are exactly symmetric.

If you put the two laws together and ask what happens when both **E** and **B** are changing, you get Maxwell's beautiful *wave equation* that says the *rate of change of the rate of change* of **E** or **B** in time is proportional to the *rate of change of the rate of change* of **E** or **B** in space.

The proportionality constant is c^2. That's all there is to Maxwell's wave equation, and you've just understood the results of a freshman physics course! And you can *calculate* the speed of light c from the extent to which vacuum magnetizes and the extent to which it electrifies.

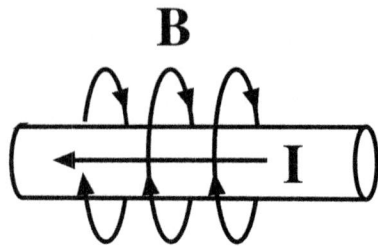

Fig. 8 B field around a wire with current **I**. **B** and **I** are always perpendicular (C. Phipps)

Radio

This whole business came about, not because of radio waves, but because telegraphers were worried about how the pulses from their keys lost their shape over long distances (Fig. 9).

You would think people would put Maxwell's equations to use right away. But it took 25 more years (1887) for someone to actually make, transmit, and receive radio waves. This someone was Heinrich Hertz, after whom we name the unit of frequency (Hertz, Megahertz etc., abbreviated Hz).

Then, lots of people jumped in: Tesla (1892), Marconi and Popov (1895), and many others you might not recognize. Marconi's system saved the people who survived the Titanic (1912).

Today, radio transmission uses a few to hundreds of MHz (Table 1). The frequencies are so high because light travels fast, and low frequencies require huge antennas to be efficient. But you can't hear millions of Hz. You can only hear 20 kHz at most, so how do we *send* voice and music or even pictures on TV? There are a lot of ways. The simplest are AM (amplitude modulation) (Fig. 10) and FM (frequency modulation) (Fig. 11). If you go into engineering, you will discover at least two dozen other ways that I won't bother you with here.

Now you need is a "detector," something that just looks at those low frequency variations and gets rid of the MHz stuff. With AM, that's a *rectifier*, something that makes a direct current (DC) signal out of an alternating current (AC) signal, plus a sluggish *amplifier* that can't follow the MHz frequencies. With FM, it's a detector that puts out less voltage when the incoming signal is off the center frequency and more when it's right on.

To better understand *rectifiers* and *amplifiers*, see *Electronics*.

You probably know you can't just go and transmit anything today. The FCC controls it. I don't want to bum you out with detail, bit I think—as a work of art—Fig. 12 is as pretty in its own way as an Arabesque [*Islamic Science*]. Or the most Byzantine thing you've ever seen. It shows you US frequency allocations today.

Radio 237

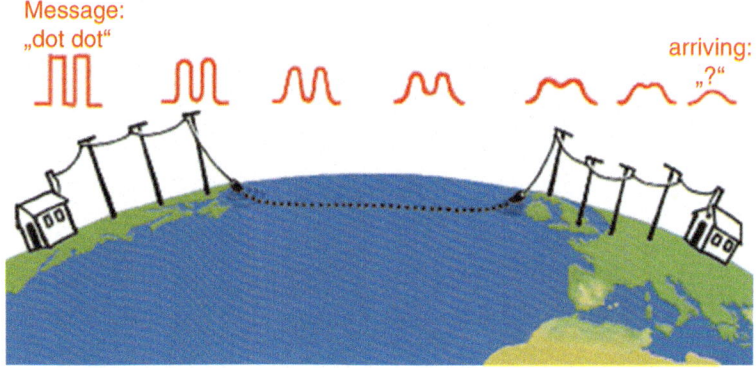

Fig. 9 Telegraphy started it all (F. Wicke)

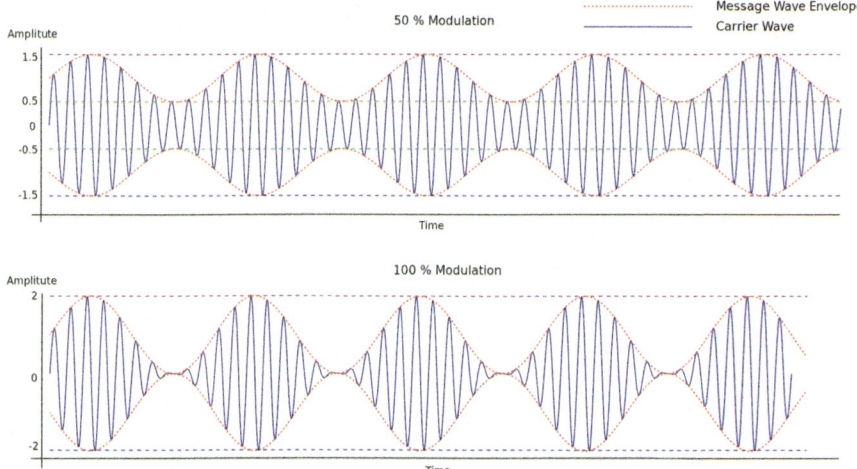

Fig. 10 Amplitude modulation (The.ever.kid Universal public domain)

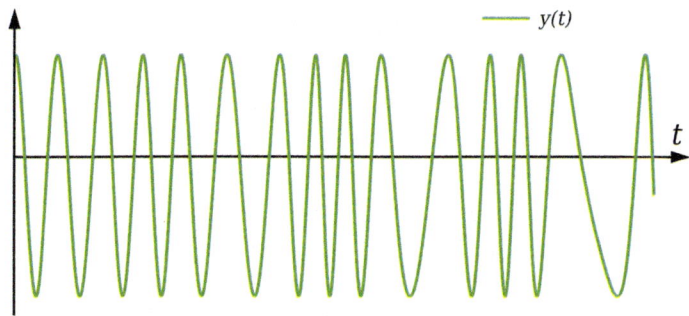

Fig. 11 Frequency modulation (Public domain. By Inductiveload via Wikimedia Commons)

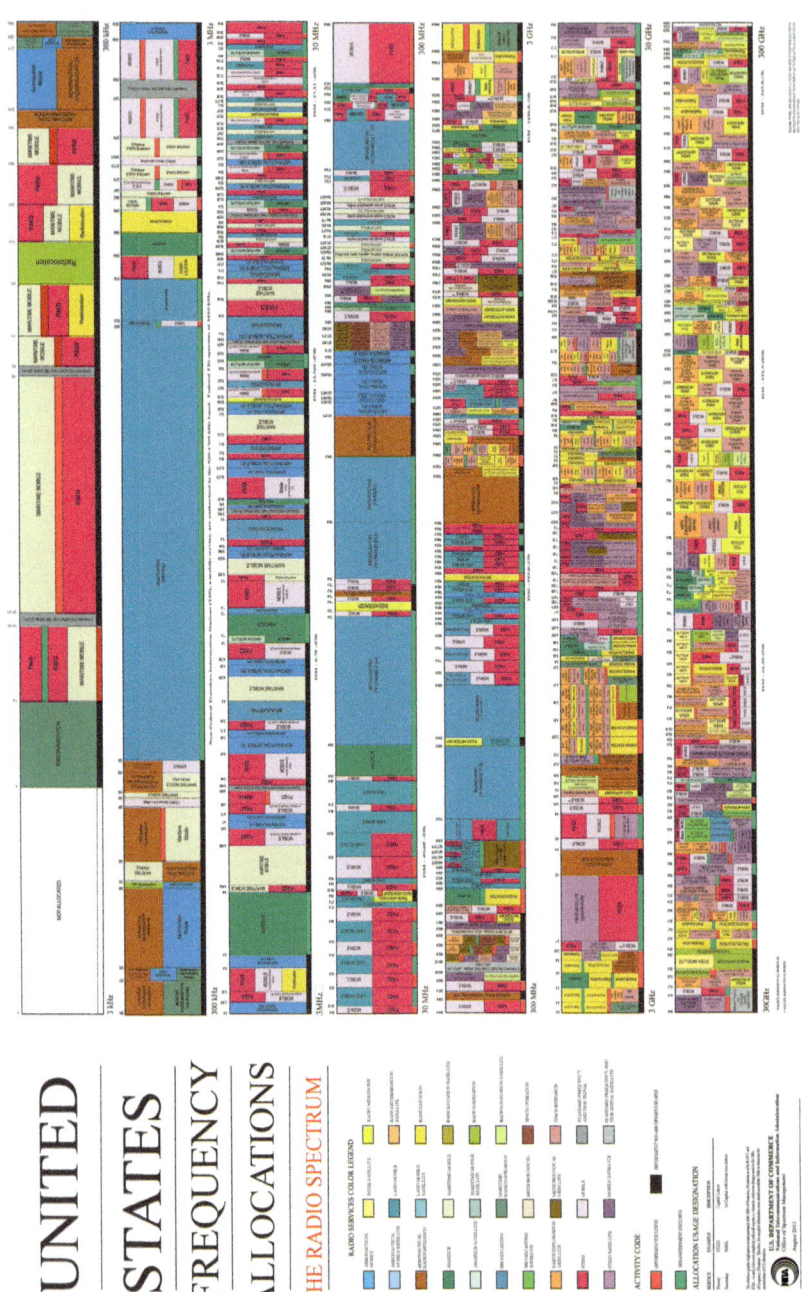

Fig. 12 The truly Byzantine regulated electromagnetic spectrum (FCC Public Domain)

Transmitters

So how do you transmit EM Waves, anyway? There are two simple ways and dozens of complex ones, of which we will mention only one.

Electric Dipoles

Any electric charge that accelerates, or any changing magnetic field, produces electromagnetic radiation. Electromagnetic information about this change travels at the speed of light. That's what Maxwell's equations tell us.

Figure 13 shows a *dipole array*, a group of dipoles on a mast on someone's roof, each length matching a different frequency. Dipoles are a quarter wavelength long, so you can see this is for a high frequency transmitter.

Magnetic Dipoles

Nobody remembers this, but in the early 1960s there was a thing called the "Landecker Ring" antenna, and I think about it every time I see this sculpture in Santa Fe, which has nothing to do with a transmitter (Fig. 14). In Fig. 15, you can see that the real thing comes from an earlier age than ours in the history of technology, reminiscent of flying saucers and such. Sorry there isn't a better picture.

This device, which was supposed to put out MW or even GW, operated by making a rapidly changing *magnetic field* rather than a rapidly changing *electric field* which is what we use today. You might guess (correctly) that the problem was the relative difficulty of making a bunch of transmitters that put out a range of frequencies, compared to the Fig. 11 setup.

Fig. 13 Dipoles (BAZ antennas via Wikimedia Commons)

Fig. 14 A Santa Fe sculpture that reminds me of Landecker Rings (C. Phipps)

Fig. 15 A Landecker Ring transmitter (US Government Public Domain)

Microwave Flutes

To make EM waves only a few cm long, you can use a sort of flute called a magnetron. The wind for the flute is a stream of electrons in a strong magnetic field. This is what is in a radar. Radars literally won World War II. They *do* put out MW and GW, and are relatively compact, so they can fit in your microwave oven.

Other versions are gyrotrons and klystrons.

To see how people make a lightwave transmitter, see *Lasers*.

Conclusion

We are literally bathed in manmade electromagnetic waves today. People are upset if they're out in the country and don't get reception. It's hard to imagine a time when there weren't any. EM waves are understood now to be a main component of the universe—from radio waves to light and cosmic rays—and we know how to transmit and receive them. We've been able to understand and use these waves for only 130 years.

Electronics and Computers

Introduction

How did we get to the Age of Intelligent Machines from the first radio transmission in 125 years? There's no one best place to start, but the amplifier is a good one. We'll be talking about big and small numbers in the first part of this chapter, so I recommend a quick review of *Numbers*.

Amplifiers

If you're 100 km away from an FM radio station that's putting out a kW of power, you're lucky to receive a nanowatt at the input to your car radio as you zoom down the freeway. That's because radio, just like sound, gets weaker the further you are from the source. It's even worse if you listen only to satellite radio, because that's coming from GEO (geosynchronous orbits) 36,000 km above your head. In order to turn that into 100 W of booming sound, you need to amplify 100 billion times.

Back in the day of the famous Titanic SOS transmission, they didn't have amplifiers, and radio was just wireless Morse telegraphy. If you could detect it, you could receive it—but you couldn't make it louder. Somebody needed to work on that, and that guy was Lee DeForest.

He invented the Audion, the first practical electronic amplifier in 1908 (Fig. 1). It's the first vacuum tube. You can understand how it works better from the electronic circuit symbol for what today is called a triode than from the picture (Fig. 2).

F is the filament, just a glowing hot wire coil operating off a few-volt battery. P is the plate. If you connect a few hundred volt battery to P and F, with P positive, you'll get a fraction of an amp of current. That's easily 100 W. Electrons are being boiled off the hot filament and going through vacuum to the plate, where they continue their journey outside the tube. The grid is a coil of very thin wire surrounding the filament. The magic is in this: As the swarm of electrons head off from F to P, it only takes a few negative volts at G to discourage them completely, or a few positive

Fig. 1 The Audion. Gregory Maxwell (GNU free4 documentation license via Wikimedia Commons)

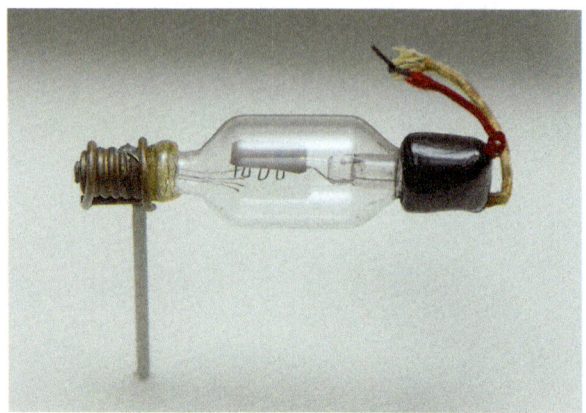

Fig. 2 The Triode's symbol says it all (C. Phipps)

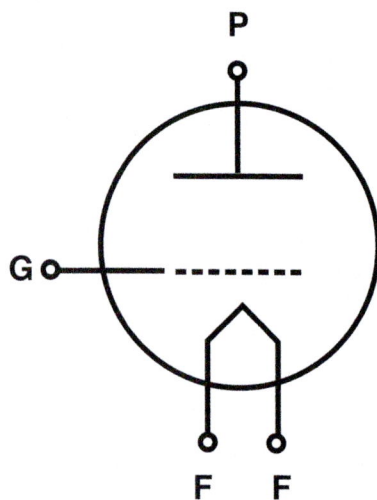

volts to encourage them dramatically. Because the grid wire is very thin, even when it's positive it doesn't absorb very many electrons. The result: with a few volts and a few mA, that is to say a few milliwatts, you are controlling 100 W. That's an amplifier—in this example with a *gain* of 10,000. Now, if you put an AC voltage at a few hundred MHz on the grid, you'll get 100 W of radio frequency power out of the tube. If you couple that power into a properly designed antenna, you're transmitting 100 W! We talked about transmitters in *Electromagnetic Waves*.

You can also imagine doing two stages of amplification, where the first tube is a *preamplifier* and the second a *power amplifier*.

For you audiophiles, vacuum tubes are also very *linear*, meaning the gain is the same at low and at high power.

Transistors, especially the FET (field effect transistor) work in a very similar way with very different physics in a semiconductor (Fig. 3). I won't get into semiconductor physics here, but you can see that D (drain) and S (source) replace P and F. G now stands for gate instead of grid, but it does the same thing, and it does it by making an

Fig. 3 The FET
(C. Phipps)

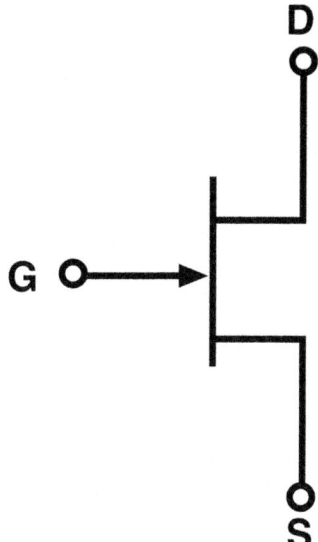

electric field so the current G needed to control the flow between S and D ranges from very small to zero and the power gain is very large. It's harder to make "solid state" amplifiers really linear, so a lot of music folks still prefer vacuum tube amplifiers.

Logic Gates: The Guts of a Computer

How logical are you? Would you like to be another Alan Turing? I hope you saw *Imitation Game*. You can skip this section if it looks frightening, but you *may* find that you have a hidden gift for computer design, even though you don't like math. They're really not the same thing.

Many different kinds of *Logic Gates* have been developed from transistors. They take two inputs A and B (which can be low or high, representing 0 or 1) and turn them into one output, either 0 or 1. Originally, these things were developed to test the truth of a complicated statement. See if you can follow Table 1. It is a "truth table." If you wish, where you see a "1," think "Yes." Where you see a "0," think "No." AND is easy. *Both* inputs have to be yes to get a yes out, because it's "A AND B." If *either* of them is a no, the output is a no. OR is also pretty easy. If either input is yes, the output is yes. NAND means "Not AND." Every place you see a yes in the AND output column, you'll get a no from this one. In the same way, NOR is "Not OR." XOR is "Exclusive OR," and a bit harder to understand but it's neat anyway. For XOR, two inputs have to be *different* to get a yes out. Table 1 is the basis of all computing logic. With combinations of these gates, you can add, subtract, multiply and divide.

Let me illustrate in Table 2 by adding two larger numbers expressed in binary (base 2) [see *Numbers*.] You add from right to left, just like in base 10. When A and B are both 1, the answer for that column is "0, carry the 1." I included a top row for

Table 1 Logic Gates

Type of Gate	Inputs		Output
	A	B	
AND	1	1	1
	1	0	0
	0	1	0
	0	0	0
OR	1	1	1
	1	0	1
	0	1	1
	0	0	0
NAND	1	1	0
	1	0	1
	0	1	1
	0	0	1
NOR	1	1	0
	1	0	0
	0	1	0
	0	0	1
XOR	1	1	0
	1	0	1
	0	1	1
	0	0	0

Table 2 Logical Addition

Carry	0	1	1	0	0	0	1	1	0
A = 51	0	0	0	1	1	0	0	1	1
B = 97	0	0	1	1	0	0	0	0	1
A + B = 148	0	1	0	0	1	0	1	0	0

the carries. In the second column from the right end, the answer row *still says* "0, carry the 1," because of the first carry. Which is why there's a 1 at the bottom of column 3.

You can do all this with AND and XOR gates. You can add using gates! If you can do that, you can add A and −B which is subtraction. There are other *algorithms* for multiplication and division with gates. And if you can do that, you've got a computer.

The First Electronic Computer

I bet you thought they used computers at Los Alamos during World War II to design The Bomb. Actually, they had a bunch of Marchant calculators and IBM punched card tabulators, and the people who operated them were called *computers*! *Punched card machines?* Check out Fig. 9 of the *China* chapter and you'll see

Fig. 4 Colossus (public domain)

a punched card machine designed in 1929 controlling a silk loom. Punched cards are a lot older than that.

The first *programmable* electronic digital computer was the British Colossus, at Bletchley Park, using 2400 vacuum tubes, completed in 1943 to help break the German Enigma code (Fig. 4). It was focused on that task and was not used for other purposes. Those of you who have seen the movie *Imitation Game* will know about this already. It had no RAM (random access memory, or just memory) at all. The output was paper tape and a printer, and the input a bunch of switches, plug panels and a paper tape reader. Its "hard drive" was paper tape with a 100 kb capacity. Colossus had an equivalent processing speed of almost 6 MHz.

Computing Speed and Parallel Processing

There are two aspects of this. Figure 5 shows my personal graph of *system speed* vs. time, in floating point operations per second or FLOPs. That top item is a Chinese supercomputer! It does 34 Petaflops, and could do 55!

While the number of transistors on a chip keeps going up [see *Lasers*], clock speed—the rate at which things happen in a central processing unit (CPU)—is stuck around 3 GHz and has been since 2003 or so. Why is that? It's because silicon gets too hot if you go faster. Back in the 1960s, I wrote a paper suggesting the answer is 3D rather than 2D circuits, but people haven't figure out how to do that yet.

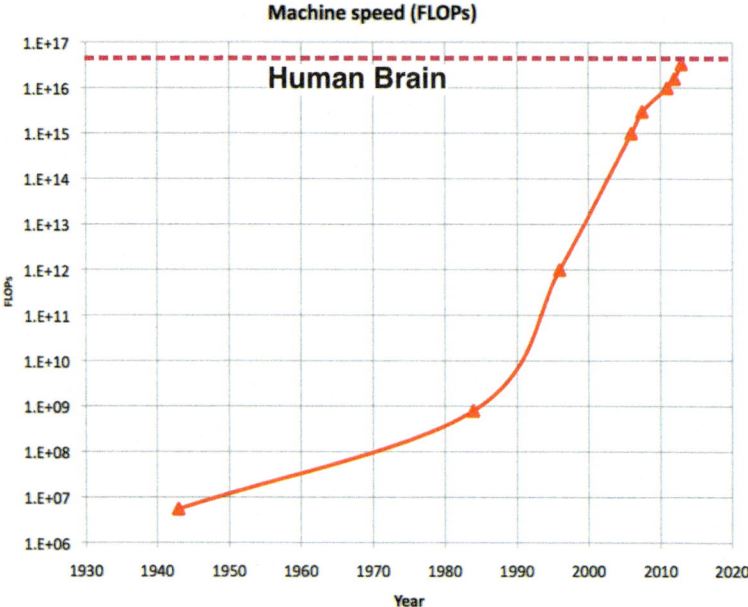

Fig. 5 FLOPs vs. time. Estimate of Human Brain capacity is from Kurzweil (see below) (C. Phipps)

Is that a problem? Not necessarily. Since about 2003, people realized that letting a whole bunch of CPUs cooperate in solving a problem works very well too. It's called *parallel processing*. Most of the recent progress shown in Fig. 5 is due to that realization.

How do you do that? Some parts of a computer program need to be done in serial fashion, one after another, but many parts can be done in parallel. If you can state the problem in such a way that 95 % of it can be done in parallel, you can do the whole thing 20 times faster.

When you bought your last Mac, did you get a dual-core or quad-core unit for better speed? This is the same idea. The Tianhe-II supercomputer in Guangzhou at the top of Fig. 5 is a *3.12 million-core machine*! Each core is an Intel Xeon CPU, working at 2.2GHz. Its memory is a million GB. That's 1 PB! Collectively, the thing uses 24MW of power! That is how you work faster. And, if you take the estimates of Ray Kurzweil seriously (see the next section), it is equivalent to a human brain.

Information Explosion and AI

Actually, I hate to call it that. People keep acting as if data is information, and it's not. Information is processed, meaningful data, and we've got far too little of that. Because we can do it, we do it, and it's getting to be difficult for one person to keep up with anything. That's part of why I'm writing this book!

I've had a long time to think about a lot of stuff, and—I am happy to admit it—the internet allows me to assemble the basic ideas in my personal memory into a

Information Explosion and AI

book that covers a lot of details, with images that I only vaguely recall, thanks to our laws about public domain and to Wikimedia Commons. I could not do this using real books in any public library that I know. So, I'm not against the explosion of data, only the lack of eagerness some people seem to have to really sort it out. Is this explosion a problem? Well, it depends.

A few years ago, Ray Kurzweil wrote a couple of books indicating the crossover of absolute computer capability and human mental capability about now, in 2015. So far, his predictions made 10 years ago are turning out nicely (Fig. 5). And, get this: because stored knowledge grows exponentially [see *exponentials*] with time like e^{at}, and processing speed does, too, you can imagine a double exponential like e^x where x is itself e^{at} for the growth of information technologies as a part of our lives.

It's great that we're so good, by the way—Kurzweil puts the human brain at 50Petaflops with a 1 TB memory capacity. Tianhe-II can do the processing speed, *today*, and has a 1000 times the memory, just as Kurzweil predicted for this date. Just processing the images as you walk or drive along takes a lot of that capacity. Especially if you're skiing through trees in the Santa Fe National Forest. That's why a turtledove jerks her head as she takes steps, taking a series of snapshots to make a low frame rate movie and reduce the data rate to what her tiny brain can handle.

Because the crossover between computer and human computing capability is already here, it will be an increasingly bigger psychological problem for folks. Problem? Look around. Already, travel agents are no more. Along with a lot of the guys that assemble cars, bookstore clerks, switchboard operators, machinists, receptionists, word processors, meter readers, typesetters, bank tellers, painters, even some surgeons and pilots—all those folks that made up a strong middle class to buy stuff we make. *It is not just Chinese competition* that's the problem! Google has cars that drive themselves. When you land on a modern flight, chances are a computer is at the controls. Middle class jobs are disappearing much faster than people can retrain for the fewer and fewer uniquely human jobs that might not be done more cheaply by a machine.

I read a story recently that predicted half of the jobs in the USA are vulnerable to "computerization." That's a lot. I was fortunate being born many years before the predicted crossover, that Kurzweil calls The Singularity. And well he might.

"AI" means artificial intelligence. Kind of an unfortunate name if you ask me. It's real intelligence—it's just not human. By AI, we mean the ability to reason, learn and store knowledge. Ever since I saw the movie *Her*, I doubted that any of this is uniquely human. You should rent it.

This is not a joke! You have to think, if you're young. You want to train and plan for a skill that is uniquely human and might last a lifetime. What are uniquely human skills? Music, Art, Physics, Mathematics, History, Creative Writing, Gourmet Cooking, Birthing, Petting a Cat, Sex! Well, OK, the last three you don't get paid for at least yet, unless you have a type of job you wouldn't really want. You need to be a high-powered practitioner of these kinds of skills, not just garden-variety. If half the folks have already lost their jobs to machines, you really need to climb to the top in some creative activity. I chose physics. My lady chose psychotherapy. Scarlet Johanson chose acting. You don't need to know math. But you need to be good.

Books are beautiful. You should visit the library, relax for an afternoon and read a book is already there, even if you don't need it for a paper or book you're writing. Each one of those books has a big piece of someone's life in it.

Is there a defining difference between humans and automatons? In the movie *Her*, the only apparent difference is that Samantha is not mobile, a physical being you can see and feel. That last piece would not be *too hard* to add to what we already have. Have you seen those videos of the Boston Dynamics robot dog, or "Cheetah" galloping along? People might already be the ultimate automatons, placed here by some interstellar Boston Dynamics. What are mind, soul and spirit as anything separate from processes in that wonderful computer in your skull? You decide!

Social Media

I'm not sure who gave it the name *Social* media, but many of us now swim in a sea of tweets, Facebook and Linked-In postings, blogs, e-mails, Instagrams, YouTube videos, texts, Google searches, and IM's that we feel we must respond to. "So-and-so shared a link!" Great. I've got stuff to do. Sorry if I sound like a dinosaur. I have read about younger folks who can't sleep through the night because they might have missed a text.

This results in a lot of internet traffic which, as we can see from Fig. 6, was doubling every 18 months in the mid-1990s. But now, it's leveling off, like CPU speed!

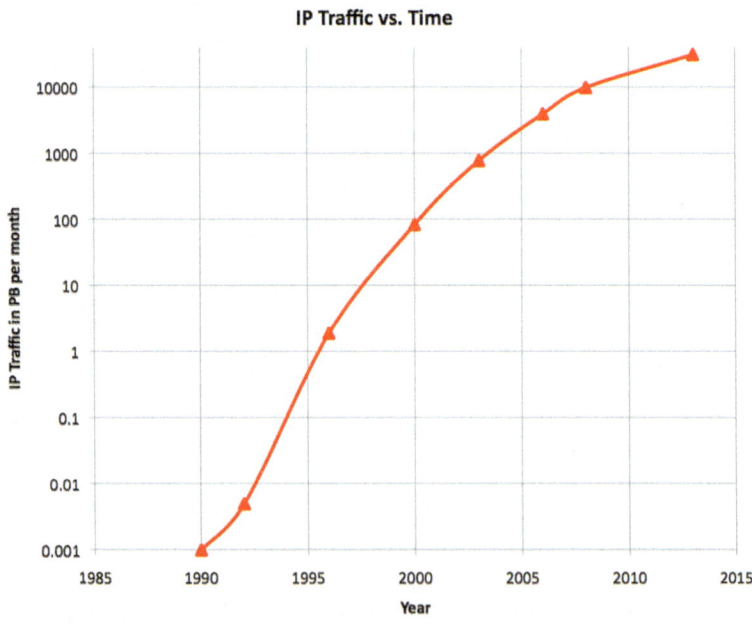

Fig. 6 IP Traffic vs. time. If half of this is from the USA, that's 1.4 GB/day for every man, woman and baby. Maybe that's why it levels off? (C. Phipps)

Not necessarily because of limits in our fiber optics, although that's part of it. If half of it is from US customers, it amounts to 1.4 GB/day for every human being in the USA, including babies. It might be getting close to as much as we are able to use!

The Universal Brain

Have you asked Google a question lately? Try "Translate: Tianhe" and you'll find it means Milky Way. Or ask it about, "I wanna hold …" By the time you get that far, it already knows you want to learn more about that Beatles song. What can I say? I depend on it. So do a lot of other people, with the result that Google uses 260 MW for its billion searches a day, and that's why their data centers are near water for cooling. I know somebody whose cousin was all excited that Google was coming to town, only to find out they just needed two employees to run the new data center, and that they moved it there because of the river. They use only 0.01 % of world power generation, but it's interesting that it's already a measurable fraction.

But note! (Fig. 7) Google searches are leveling off, too! When I looked into that, I expected to see an exponential.

World Power Generation? There's a favorite subject! About 2.5 TW (million million watts) last year. Did you know that amount of power is only *3.5 times larger than what the 7.1 billion human beings on the planet put out by just living?*

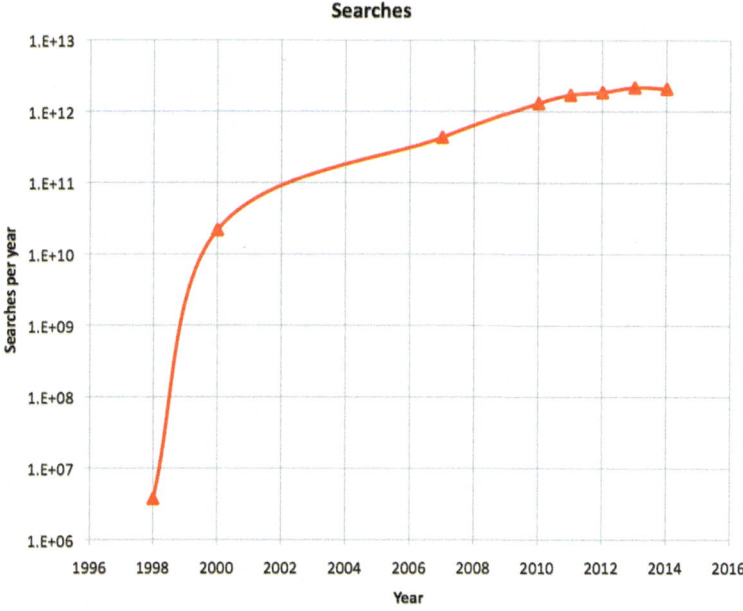

Fig. 7 Google searches per year. It's leveling off, too! (C. Phipps)

iDevices

If there's one man who changed the entire world since I've been alive, it was Steven Jobs. Look at Fig. 1 of *Optics*. Those young ladies are proof that iPhones have spread across the world. This is a device no one knew they needed until 2007. Seven years later, Apple sold its 500 millionth iPhone. All told, 2.8 billion smartphones have been sold, two for every family on the planet. In Stockholm recently, I noticed a father pausing behind a baby carriage for a moment to consult his phone and thought, "Wow, this is a lonesome way for a child to grow up!" until I noticed the child was fingering his phone, too! What can I say? It's a different world, and the connectivity is infinite. When you have that, *anything can happen*!

Robots and Drones

There was a novel titled *VOR* by James Blish back in the 1970s. It frightened me. In the far-distant future, a fearsome robot arrives from a distant planet and wreaks havoc for a while. And many of us also remember *The Day the Earth Stood Still*, featuring Gort. He and his handler Clatoo have come in peace but, if you don't obey him … Zap!!

Well, it's happening! Boston Dynamics, Schaft … several companies are working on pretty capable robots already (Fig. 8). Back in the Iraq War, cruise missiles

Fig. 8 Boston Dynamics' Cheetah. Yes, developed for the Army (DARPA Public domain)

were finding their targets by themselves and ICBMs are not entirely dumb. Today, when you send a space probe to study a distant comet—like the European Space Agency's very capable *Rosetta*, it needs to be robotic. That comet is 488 million km away from Earth right now, so it takes light almost an hour to make a round trip. It would be silly to try to control its landing on the comet from here.

A robot is an intelligent machine that recognizes its environment and functions with only a minimum of external commands.

Remotely controlled drones like we have now are one thing, but *robotic* drones are quite another. I just read that Amazon might want to deliver things that way. We talked about that in *Drones*.

"Artificial" Beings and Neural Nets

Of course, like Artificial Intelligence, they might be very real. Go to YouTube and watch one of Boston Dynamics' beasts come to life, test itself and start galloping across the ice. Pretty it up, make it stand and combine it with the personality of Samantha in *Her*. What have you got? A pretty convincing being.

You may have noticed that your brain is not running on a linear, line by line computer program. It is instead a massively parallel *neural net*. Neural nets are designed to get *good enough* answers to questions involving a whole bunch of simultaneous inputs by learning through teaching, trial and error. An example is *pattern recognition*, one of the skills those cruise missiles had. Even though Scarlet Johanson is not a neural net per se, the skills she demonstrated in the script for *Her* are specifically the ones artificial beings will have to seem *totally human*.

Kurzweil predicts "The *Singularity*" by 2045. By this he means that AI will be so good that people envy machines and want to upload themselves into one. I doubt that, although I could be wrong. After the upload, *you* have to be killed, right? That's a pretty hard sell.

Why does it take 30 more years, considering what's happened in the last 20? Tianhe-II covers an area like a basketball court, uses 24 MW and is a *very* long way from moving around on two feet. It doesn't have to move around to be a competitor but, it helps (Fig. 9). This young lady robot was developed for customer service at the 2005 Expo Aichi in Japan. She answers questions in Chinese, Japanese, Korean and English. Does this bother you?? What about this one?? (Fig. 10). Be prepared!

To make it *really* clear how I feel about education, I offer you Fig. 11 ….

Stephen Hawking recently warned "… computers double their speed and memory capacity every 18 months. The risk is that computers develop intelligence and take over. Humans, who are limited by slow biological evolution, couldn't compete, and would be superseded." That's the worst possibility.

Fig. 9 Actroid (Gnsin Creative Commons license via Wikimedia Commons)

Fig. 10 TOPIO (Humanrobo Creative Commons license via Wikimedia Commons)

Fig. 11 PEARLS BEFORE SWINE © 2015 Stephan Pastis (Reprinted by permission of Universal Uclick for UFS. All rights reserved)

Conclusion

The science of electronics produced long-range communication devices and then computers, without which we would be in trouble today. *Computers* today are mostly designed to be rigidly logical. Very soon, that will not be so. In 10–20 years or so, *Samantha* will not be a fantasy figure, but may instead be your boss. More and more jobs that can be automated already are. Get ready!

Next: the stunning field of Biology!

Biology

DNA (deoxyribonucleic acid) is the code of life. It is a base-4 coding system, using molecules called guanine (G), thymine (T), cytosine (C), and adenine (A), and the human genome contains about 1.5 GB of information, 50 % more than my copy of Microsoft Office, at least. *Each cell* in your body has essentially the same information about how to make you! Can you imagine that, considering how small a cell is compared to an Office Installation DVD? But yet, the instructions for *you* are not more complicated than Microsoft Office.

You wouldn't think it when you stretch the slimy stuff, but a single DNA fiber is four times stronger than structural steel!

Watson and Crick (with major help from a young female scientist, Rosalind Franklin) figured out the unusual shape, and got a Nobel Prize for it in 1962. As you can see, RNA has uracil (U) instead of thymine. It can be thought of as the original copy of you, and DNA the working copy. DNA acts through proteins that it causes to be made in your cells with the help of RNA. RNA can tell DNA which traits to switch on or off (Fig. 1).

Today, we can actually see these things with electron microscopes [just 3 years ago—see Optics!!]. But in the 1950s, it was Franklin's X ray diffraction patterns that gave those two guys the clue about the "double helix." Nobel winners have to be living, and she was dead.

Mendel

In 1856 (so recently!), a Moravian priest called Gregor Mendel started growing peas in his garden and recording how successive generations passed on qualities like plant height, shape, and color. The offspring of yellow and green peas was always yellow. But in the next generation, green peas were ¼ of the crop. He invented the idea of recessive and dominant genetic traits to explain what he saw.

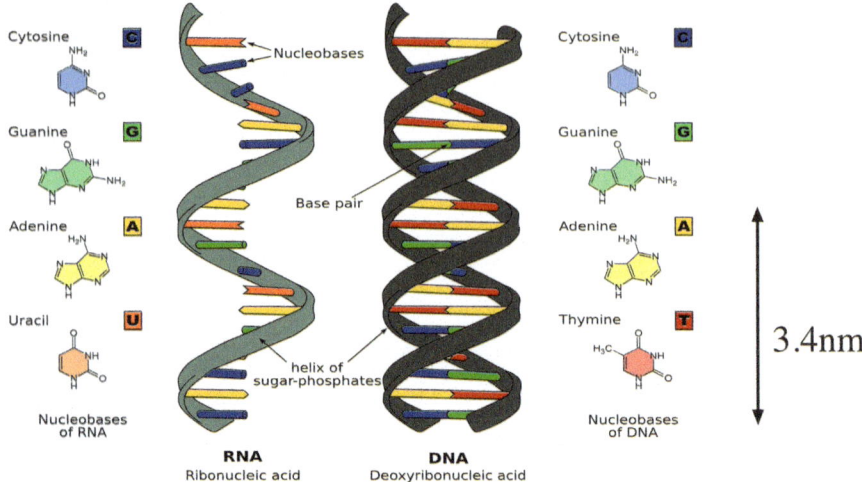

Fig. 1 DNA and RNA (http://upload.wikimedia.org/wikipedia/commons/3/37/Difference_DNA_RNAEN.svg Sponk GNU Creative Commons license via Wikimedia Commons)

Darwin

Three years later, the British scientist Darwin published *On the Origin of Species*, undoubtedly the most important book in the history of biology. Twenty years before that, he had sailed all around the world observing animals and collecting fossils. He brought home lots of samples, but not tortoises—he ate those instead. What he found was that accidental changes in genes that enable a species to compete better will enable it to leave more offspring. This is the *theory of natural selection*. It is also *evolution*. His book sold out. Simply stated: things evolve capabilities in order to survive, and different species have had to survive different threats.

In other words, species can change!

Next was *The Descent of Man* which claimed that men descended by evolution from apelike creatures. This was as heretical at the time as the idea that the Earth goes around the Sun. Even in the late nineteenth century, public reaction was not all positive (Fig. 2). In many quarters, the idea is still heretical today.

Nobel's brother Emil had only recently blown himself up, and it would be another 37 years before the first Nobel Prize. Darwin's reward is burial in Westminster Abbey.

"Sequencing" Human DNA

In 2003, the world's most ambitious biological project was complete—actually determining and writing down the human DNA code. Using technology available then, it took 13 years and cost $3 billion. This project got going right here in Santa Fe. Whose DNA did they sequence? It's a secret. But a lot of it came from some guy in Buffalo, NY. In 2007, the sequence was published.

Fig. 2 Darwin the ape, 1871 (Public domain)

Twelve years later, this is a multibillion dollar industry, and you can get yours done in a few hours for $1000. You can buy the machine that does it for $700k. How is it done? First, you cut it up in little pieces, then feed it through a machine.

How does the machine work? In 1970 it was very tedious, but today you can cut DNA into little pieces, each one word shorter than the next, and make them glow different colors in a laser beam, depending on what word is on the end. Then you can imagine squirting that stuff through a very fine pipette and just asking the computer to read the colors.

So how do you cut DNA?

Snipping Genes

Aha! The tool is called molecular scissors, or "restriction enzymes." They are able to recognize a particular sequence and cut *right there*. Strep bacteria developed this defense against viruses billions of years ago. Wow! And today we *use their invention*!

This process was first understood and done deliberately in 1970.

Today, artificial enzymes are used, with names like "zinc finger nucleases" and "transcription activator-like effector nucleases"… Oh heck—We'd both have to have a Ph.D. in biochemistry to understand the details. Figure 3 shows such restriction enzymes working. They really are clamshell shaped proteins that act like scissors!

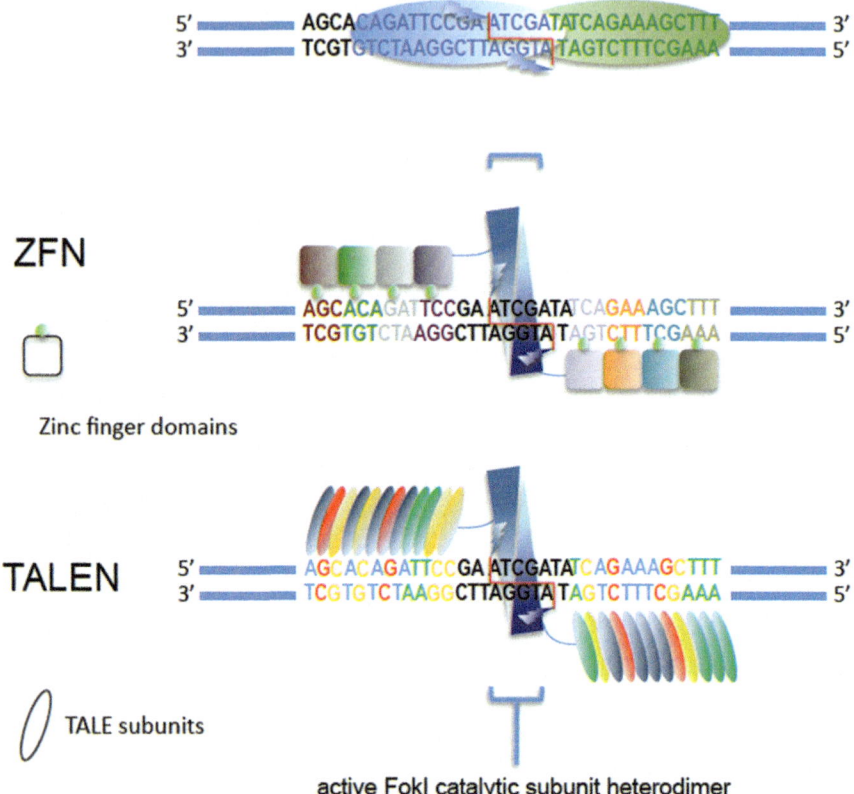

Fig. 3 Molecular scissors (Farzad Jamshidi, GNU Free documentation license via Wikimedia Commons)

Reassembly: Modern GMO vs. What We've Done All Along

Synthesizing DNA has been done since 1983 with the polymerase chain reaction for which Kary Mullis got a Nobel Prize. This is just a way of making millions of copies of a particular DNA sequence. That's important to get enough of it to study and use. It's also a key tool in cloning.

You can clone cells and you can clone entire organisms, like Dolly the sheep in 1996.

Word-by-word manufacture of DNA has also been done since 1972, when people first made yeast genes from scratch.

Fig. 4 Glofish
www.glofish.com

You can splice plant DNA to bacterial DNA, or human DNA with DNA from a fungus. That's *recombinant DNA*. You can make a fluorescent fish (Fig. 4). You can make human insulin without humans. It's the only way to make hepatitis B vaccine.

In the USA, believe it or not, you could be eating a cloned sheep without knowing it—the FDA does not require specific labeling.

We have all heard of GMO corn, developed to tolerate a specific herbicide.

When a food organism like corn can be artificially designed to have certain other qualities, like producing an insecticide, one might wonder about the "trust me" aspect. "What can it do to me?" is a natural question.

As a general procedure, human dominated selective reproduction has been done for tens of thousands of years. That's how a scrawny wheatlike thing became corn in the first place. Corn as we know it is not natural (Fig. 5).

But these techniques have been imprecise and inefficient. Now, a new technique has changed that.

Fig. 5 Evolution of corn, a GMO by human controlled natural selection (John Doebley, Creative Commons license via Wikimedia Commons)

Rewriting DNA Code

We can now *edit* DNA in a way that can be inherited. Jennifer Doudna at UC Berkeley and Emmanuelle Charpentier of the Hannover Medical School are major figures in this story. It turns out that a special protein called CRISPR/Cas9 could be programmed to cut precisely any piece of DNA from any organism, starting at one specific place and ending at another, and do it quickly and accurately. Then you can glue pieces together and put these—ummm *transgenic* pieces into cells and make them reproduce. You could repair, or enhance, any human gene. Maybe improve intelligence or beauty.

Not everyone trusts scientists to make decisions for all of us. I am reminded of Edward Teller, in 1946, making a quick calculation on the world's behalf that concluded the first hydrogen bomb would *not* initiate a chain reaction in the atmosphere. So, he went ahead with testing it. Well, it did turn out OK!

Reasonably, Doudna called for a temporary moratorium on using this technique until people understand it better.

On this, Stephen Hawking is more sanguine than he is about AI. "Humans have entered a new stage of evolution," he says, adding "I think it is legitimate to take a broader view, and include externally transmitted information" [meaning what you can think up] "as well as DNA, in the evolution of the human race."

Does this remind you of how things multiply in *Exponentials*?!

Conclusion

Biological science is the most rapidly changing scientific field at the moment, producing fundamentally new tools every 2–3 years. Even laser physics doesn't change that fast. DNA engineering offers new ways of understanding and curing diseases as well as for optimizing qualities of living beings much faster than Nature herself. Just like some developments in physics, this is a double-edged sword.

Next: the fascinating fields of Optics and Lasers.

Optics

In *Islamic Science*, we talked a bit about the design of *lenses* and *mirrors*. In *Lasers*, we mention how a perfect *lens* can turn 1 W into 100 MW/cm^2, and we will show a spectrum, which comes from a spectrometer, which originally used a *prism* and now uses a *diffraction grating*. These are examples of *optics*. The way light works as optics transform it is the science of optics, which has produced telescopes, microscopes, lasers, cameras, spy satellites, optical fibers, and Real3D. Not to mention Bose–Einstein condensates in which light travels at the speed of a bicycle.

Mirrors: The First Selfie

Just 8 years ago you would have said "what the hell are these people doing?" (Fig. 1). Eight hundred years ago, aside from being judgmental about their dress, you would think "That's a mirror on a stick." Today, you instantly register "Selfie!" Steve Jobs is smiling.

There's no mystery about what a flat mirror is and how you use it. It lets you see yourself. That's an important feature for shaving and applying makeup, so as far back as 6000BC people had polished obsidian mirrors. China had bronze mirrors in 2000 BC, and the Romans had "silvered" glass mirrors coated with gold leaf.

Signaling in the Raj

A century and more ago [see *Rome* about ancient signaling!], British soldiers in the Raj (India) used mirrors very effectively as signal devices. In fact, *how* they used them was very clever. The principle is perfect for keeping solar collectors pointed at the sun. Here's how it worked (Fig. 2). A soldier holds a flat, half-silvered mirror with a "+" scratched in the silver, and looks through it to the point where he would

Fig. 1 Selfie (Petar Milošević by written permission)

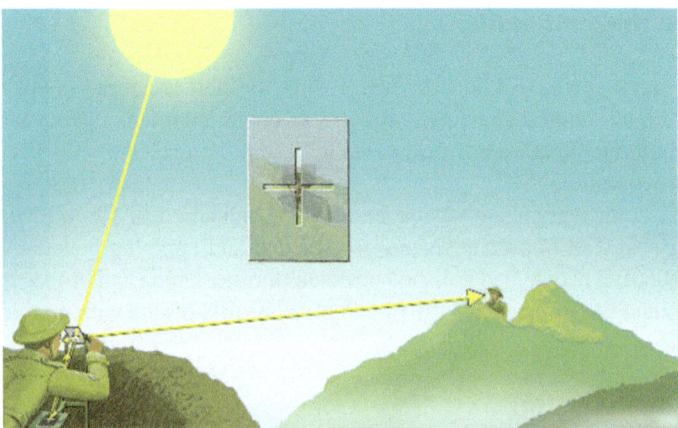

Fig. 2 Signaling in the Raj. It doesn't have to be noon (C. Phipps/F. Wicke)

like to send a flash. He can also see the reflection of the bright "+" on the ground. When he aligns that reflection perfectly with his target, he's aligned!

Think about it until you understand it! This is a lesson in optical logic. Now he can turn it away and back at will, sending signals in Morse code. Replace the eye with a semiconductor quad sensor and you've got a sun tracker!

Curved Mirrors

There are two ways to make an image (Fig. 3). I think you know that. Partially transmitting curved and flat mirrors are important laser components because you usually need to pick off a part, or even most, of the output and send it back through the "gain medium" to get the light to amplify instead of dying out.

A mirror shaped like the inside of a sphere will work just like a lens to focus light. That guy projecting an image of himself in Fig. 5 of *Islamic Science* must have been holding a concave mirror. A lens is easier to use, but glass is opaque at certain wavelengths—UV and X-rays for instance—and you *have* to use a mirror. A convex mirror (shaped like the outside of a sphere) lets you see around corners in the supermarket, or on the curve in the road at my driveway so I see what's coming.

Lenses: The First Optics

The Assyrians had lenses made of rock crystal in 700 BC.

As we get older, the adjustable lens in the eye stiffens and its focal *range* decreases. Ibn Sahl and Al Haytham (*Islamic Science*) were probably the first to make corrective lenses that people could wear to correct vision problems. They understood lenses well, but Al-Haytham's *Camera Obscura* was probably just a dark room with a hole in the wall. This is a version of the pinhole camera, which has the beauty of making a clear image of everything, far and near (Fig. 4).

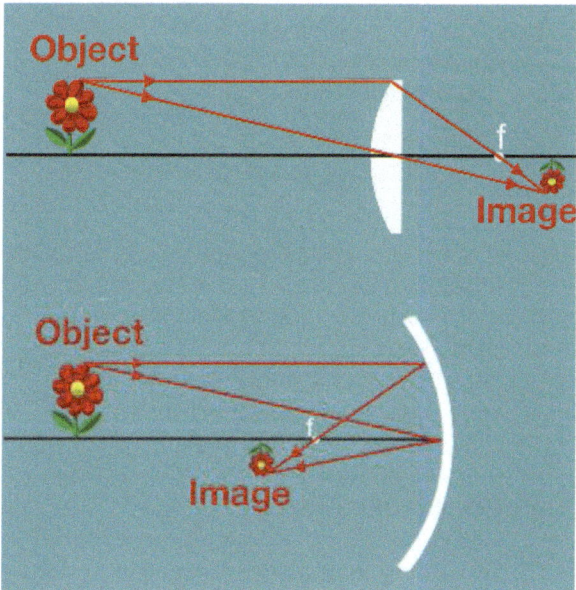

Fig. 3 How lenses or mirrors make an image. "f" is the point where light from infinity would focus. Light from a closer object at "O," comes to a good focus at "Image," which is farther than "f." You can see why the image is upside down (F. Wicke)

Fig. 4 *Camera Obscura*. Some people think artists used this device to help them paint (Public domain)

The neat thing about lenses is that they make an image of the whole world you can see. A real camera, like your eye, lets in a lot more light than a pinhole, but the trade-off is that the more light it lets in, the more it only makes sharp images at one distance. Figure 3 shows why this is so for lenses or concave mirrors, and why the image is upside down.

Lens design is a complicated business, as you saw in *Islamic Science*. People have been doing it for hundreds of years.

Cameras

When people got tired of being in a dark room with a tiny hole in the wall to make images of the world, they started thinking about how to let in more light while keeping a sharp image, and they naturally thought about how our eyes work. Make the hole bigger and put a lens in it!

Now, how to take a picture with something portable, and keep it? There's only room to mention the really key inventions in this chapter.

Black and White Photos and Movies

Louis Daguerre figured out how to make photos on copper coated with silver and silver iodide in the early 1800s with a barely portable camera with a lens in the end. But, *Daguerrotypes* took several minutes to expose.

Fig. 5 Kodak Brownie Camera (Håkan Svensson GNU Free documentation license)

What we're talking about here is *photochemistry*, finding a way for light to make changes in a surface that you can later *develop* to make them really visible, and then *fix* them so they won't change in the light you're using to look at the image. Until videocameras and iPhones, most of those processes used silver, platinum or palladium chemistry. Even today, photographic artists use platinum printing for the very best results.

But—people wanted faster photos! George Eastman built an empire out of dry plate photography, superseding the wet gelatin process of the 1870s. By the late 1800s, Eastman Kodak was selling flexible transparent film in rolls as well as a camera to use them (Fig. 5). These, of course, were all black and white. It was then a short step to 16 mm black and white movies in 1923 then color 8 mm film for home movies in 1935 [see "Kodachrome," below].

Chester Carlson

I can't help mentioning Chester Carlson and the xerox here. He wanted to copy *documents*. You can take a picture of one and go look at it, but it's another thing to print that picture by the hundreds on paper and have it look like the original. In 1938, he found that melted sulfur on a zinc plate would hold a charge, that he could discharge that with light, making an electrostatic image, which could then pick up lycopodium powder which he then rolled onto ordinary paper to make a permanent image! It was a classic mad-inventor-in-the-kitchen story. It took 10 more years working with research groups at Haloid and Battelle to perfect the process of *xerography* (combining Greek words for dry copying). But, at the 1948 meeting of the Optical Society of America, he could demonstrate dry copying on paper at 1200 ft per minute. Finally, in 1959—21 years and tens of millions more research dollars later, the Xerox Corporation came out with the 914 copier, so named because it made copies on

9 × 14 in. paper. It no longer used sulfur, of course, but a semiconductor drum and made copies at four cents each. The only problem was its 300 kg weight, but they leased it and guaranteed repairs. Carbon paper was out! Everybody had to have a xerox! People began copying parts of themselves. By 1970, there were 100,000 employees at Xerox and 1.7B$ in revenue, 200 times the research investment.

Kodachrome!

Back in the 1850s, the scientist Jim Maxwell (mentioned in *EM Waves*) realized that you could make color images by making three black and white photos with red, green and blue filters and then projecting the result through the same filters, carefully overlapped on the screen. Various versions of this sort of thing were done, but the history is interesting only to a historian.

By 1940, you could take a bunch of Kodachrome slides on a single roll of film, order color prints from Kodak, and make 16- or 8-mm color movies (Fig. 6). And the Kodak empire continued to grow. It was pretty expensive compared to black and white, but people enjoyed having pictures in full color.

Another whole branch of color photography is Technicolor, beginning in 1916 and probably coming into your awareness in films like 1938s *Gone with the Wind*.

Fig. 6 Bell and Howell 8 mm movie projector (Aka Kath Creative Commons license via Wikimedia Commons)

This was an incredibly complex process involving three rolls of film exposed through red, green, and blue filters, dyed and recombined into a single strip. Filmmakers hung onto that technology because it gave more vibrant color, and the color didn't fade with time.

Polaroid

Edwin Land made instant photography possible with the Polaroid Land Camera in 1948 (Fig. 7). Actually, the name "Polaroid" comes from the polarizing plastic film he invented and that we still use in Polaroid glasses. But his fame comes from instant black and white photos and then instant color. I used those at my wedding.

But here's a weird phenomenon that people still don't totally understand! Vision *really happens in the brain*! In the late 1950s, while fiddling with color photography, Land accidentally discovered that he could combine a red image with a white light image that had been recorded through a green filter and give the sensation of full color! Land never did find a way to utilize this discovery, but I can vouch for it—I saw him demonstrate it at MIT.

3D

Now here's the final amazing story in this section on cameras—how we have realistic 3D movies in color today. To get 3D, you need to take two pictures from different points of view and get those two images to your two eyes separately. The old stereo viewers did that by putting the two images in front of lenses in a special viewer you put in front of your eyes.

Fig. 7 Polaroid camera (Cburnett GNU free documentation license via Wikimedia Commons)

Fig. 8 Charles Street, Boston, late 1800s (John Soule public domain)

You don't really have to have a viewer. Relax and sit in front of Fig. 8, about 15 in. back. Now, cross your eyes enough that the two images overlap, while trying desperately to keep each eye focused. Be patient—it takes a minute. Look in the middle between the two images. Try harder and keep your eyes level. Ah! There you are on Charles Street more than a century ago! Something about 3D adds a lot of reality. Your brain does the combining. All you have to do is get your two eyes to each pay attention to different images!

You'd get tired doing this for a long time. For black and white 3D movies like *Creature from the Black Lagoon* they combined two rolls of black and white film taken from different positions through red and blue filters in a single film, and put red and blue glasses on the viewers. Your brain eventually sees not color but 3D B&W. And it doesn't matter how you tilt your head.

Now how do you get 3D color? You wear *Polaroid* glasses with one vertical and one horizontal polarizer over your right and left eyes, looking at a scene coming from two projectors projecting color film through the same polarizers onto the screen. The polarizers are gray, just like sunglasses, so they don't add any color and you see 3D color. But wait! The minute you tilt your head, the images get mixed up! That really makes your neck sore after an hour or two. And installing and perfectly synchronizing two projectors is a drag!

The solution? Remember RCP and LCP light from *EM Waves*? Here is one reason I went through that. RCP and LCP really are the elements of light. You can do the same 3D trick with RCP and LCP glasses in front of your eyes, and now you can tilt your head. In *Real3D, one* digital projector sends left- and right-eye color film scenes that are alternately RCP and LCP to the screen. They do that 144 times a second and, with the help of those glasses, your brain does the rest. Isn't technology wonderful? I really thought Real3D was a miracle when I first saw it (*Avatar*).

Who invented that? Lenny Lipton, the same guy that wrote *Puff the Magic Dragon* when he was still in college! Don't you wish you did?

Prisms and Spectrometers

Prisms are fascinating. They can bend light in your binoculars like good mirrors, or be the heart of a spectrometer. As you go from red to blue, the refractive index of glass gets larger, so blue light gets bent more than red as it goes through the prisms in Fig. 9. In 1665, the brilliant Isaac Newton showed that white light can be broken into a rainbow of colors. More important, he showed he could reassemble the same spectrum into white light. This was stunning at the time. Before that, folks thought white light was pure and colorless and that a prism somehow messed up the purity.

It's the same thing in a rainbow, of course—after two reflections in a rain drop—the rays always go at the same angle with color dispersion because of refractive index differences in *water* (Fig. 10). The result is a rainbow! (Fig. 11)

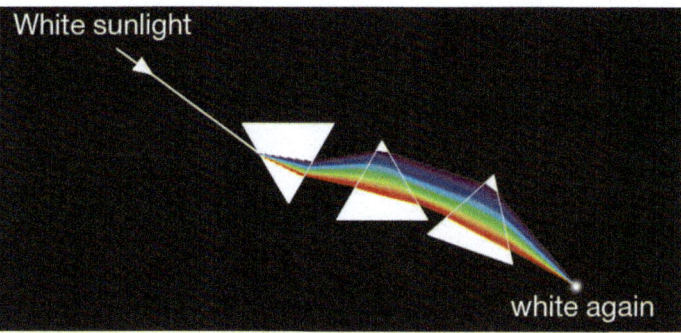

Fig. 9 Prisms and spectra. Newton's great discovery was that the colors can be reassembled into white light (F. Wicke)

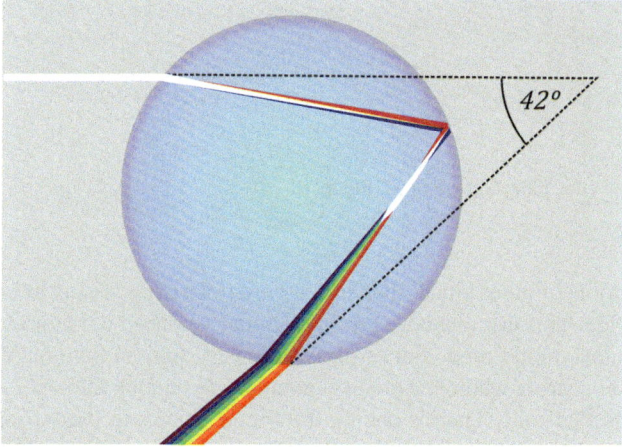

Fig. 10 Raindrop reflections (KES47 public domain via Wikimedia Commons)

Fig. 11 The rainbow. The shadow of the photographer's head is the center of the circle (Eric Rolph at English Wikipedia Creative Commons license via Wikimedia Commons)

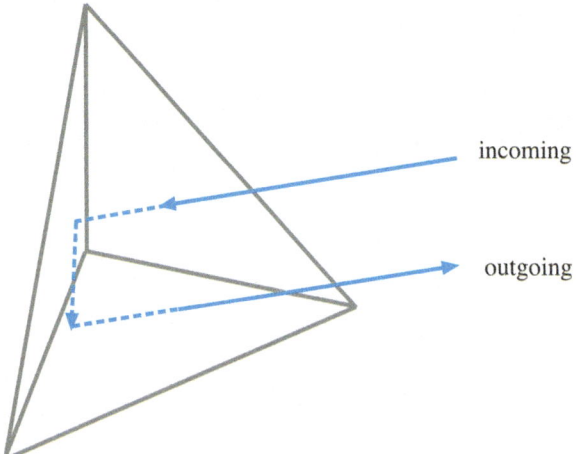

Fig. 12 Corner cube reflecting prism (C. Phipps)

There are lots of other kinds of prisms—from polarizing prisms to a corner cube prism (Fig. 12). This last one is *so* fascinating that you need to know about it.

This is just a corner made out of glass. It always returns light back to where it came from, not caring where the prism is pointed. If you look into one, you see your own eye, and that's all! There's one on the Moon, thanks to the astronauts, and if you know roughly where it is you can flash it with a laser and get a flash back and measure the distance from how long that trip takes (a little more than 2 s).

Corner cubes don't have to work with light. In your own room, if you pound a pingpong ball into a ceiling corner, it will come back at you. In a room or stairway with a concrete or marble corner, you can clap your hands and get a sharp echo. With radar, a metal corner works the same way.

Diffraction Gratings

Diffraction is what happens when light passes a sharp edge. A grating is what it sounds like, a multitude of fine metal slits on a piece of glass (transmission grating) or many fine grooves on a shiny surface (reflection grating). When the spacing is as small as the wavelength of light, you see colors.

You have a reflection grating in your CD or DVD collection. Shine a green pointer laser on the rainbowy side and you'll see a bunch of reflections. One will be in the direction where you would expect it to be if it were a mirror. One will be way over where it shouldn't be in the reflection direction. One, two, three or more will come back toward the hand you're holding the laser with, which would never happen with a mirror! This is what happens when the grooves sync with one, two or more wavelengths. Be careful with your eyes.

Notice this (Fig. 13): blue is bent more *refracting* through a prism but less *diffracting* off a grating.

Spectrometer

Anyway, if you bring any light in a parallel beam like the laser beam onto your grating or prism and record where those reflections go (with film in the old days) you have a spectrometer.

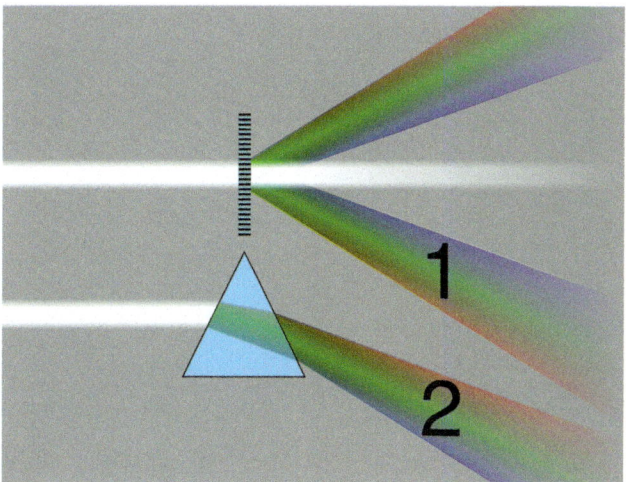

Fig. 13 Transmission grating and prism (C. Phipps)

Telescopes

Four centuries ago, people realized that a short focal length lens in front of your eye combined with a long focal length lens at the other end of a tube of the right length can make far objects look big. Vice versa, if you turn it around, the same instrument can make near things look big. The magnification factor is the ratio of the focal lengths.

You can use curved mirrors instead of lenses, and then you don't worry about focal length changing with color (Fig. 14).

Visible and Infrared Telescopes

How far can you take the telescope idea? Very far indeed. On Earth, people have made telescopes with big ("primary") mirrors up to 10 m diameter (Keck, in Hawaii) and a Thirty Meter Telescope will be operating in 2022, also in Hawaii. Not surprisingly, it's called the TMT (Fig. 15).

In space, we all know about the Hubble, an 11-t satellite 550 km above the Earth, circling us every 96 min. It's been sending back stunning images since 1990.

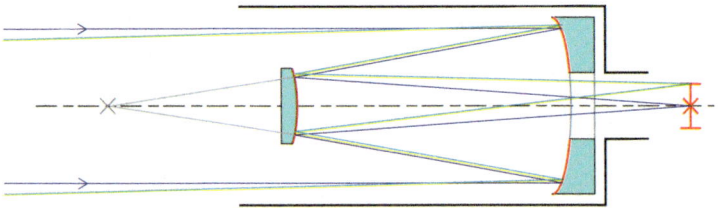

Fig. 14 Reflector telescope (Krishnavedala, public domain, Wikimedia Commons)

Fig. 15 Artist's rendition of the Thirty Meter Telescope (Courtesy TMT Observatory Corporation, via Wikimedia Commons)

Its primary mirror is only 2.4 m diameter, but, free of the smoke, haze and twinkle of the atmosphere, it sees much better than larger telescopes on Earth.

The TMT will work in the invisible infrared and have 140 times the light gathering power and ten times better vision than Hubble. Its focal length will be 0.4 km! it will be studying dark energy, dark matter, black holes and light from *individual* stars in galaxies out to 300 Zm (30 M light-years!).

In space, the James Webb telescope will have a 6.5 m mirror and we hope it is launched in 2018. With all the austerity, it's not a sure thing. It will see from the visible out to 29 μm in the infrared. In space, it can do that, but on the ground it could not. And it can see through dust in space.

X-ray Telescopes

These have to be in space. Chandra (named after the great Indian scientist Chandrasekhar) is the best example (Fig. 16). It was planned to work for 5 years, but it's still up there doing its job 16 years after its 1999 launch. This 22-t object orbits from 16 to 133 Mm above Earth. It sees from 10 nm to 100 pm.

Why on Earth would you want to make an X-ray telescope? Go back and take a look at Fig. 28 of *Weird Reality*! There's a snapshot of the black hole at the center of our Milky Way galaxy, and you couldn't see it any other way.

More recently, NuStar was put in orbit and is able to see ten times shorter wavelengths than Chandra.

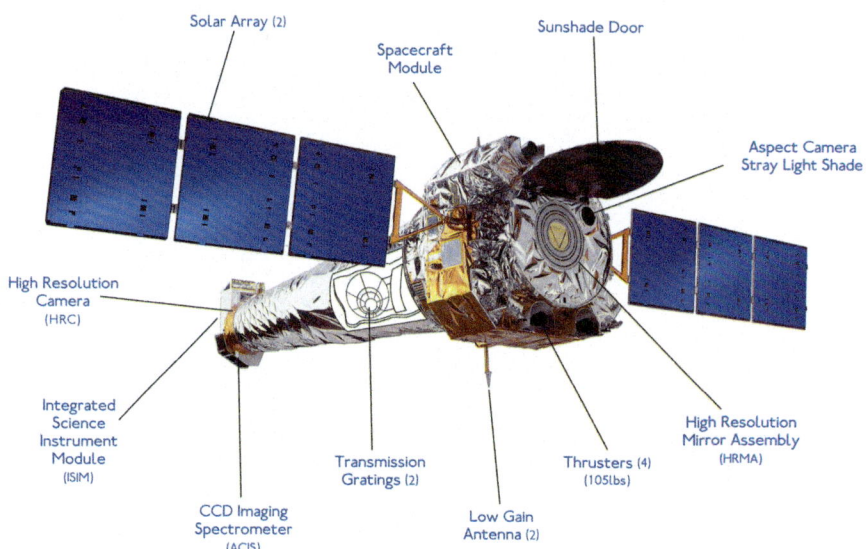

Fig. 16 Chandra X-ray telescope (NASA public domain)

Gamma Ray Telescopes

Really? Yes! They exist too! The Fermi Telescope has been working since 2008, and sees down to 4am. Not 4AM but 4 attometers! What can you use for a lens? Not much. Still, its vision is about as clear as your own. It can see where mysterious gamma-ray bursts are coming from instead of just detecting that they happened.

Microwave and Radio Telescopes

On the other extreme, the Planck telescope which went bellyup 3 years ago, saw from 350 μm to 10 mm. Radio telescopes work down to 1 m wavelength. Figure 17 shows the huge Arecibo Telescope in Puerto Rico.

Microscopes

Electron Microscopes

You probably know about microscopes from Biology. These can see about as small as the wavelength of light they use. In *Lasers*, we digress briefly into a discussion of semiconductor Fabs, and mention that you can get down to 50 nm resolution with short wavelength light. Here, we'll only talk about electron microscopes.

Fig. 17 Arecibo radio telescope, 1 km diameter. It mainly points via the Earth's rotation, but the receiver suspended over it can be moved back and forth a few degrees (National Astronomy and Ionosphere Center, Cornell U., NSF [Public domain], via Wikimedia Commons)

Crystallographic Image Processing

α-Ti$_2$Se: *Pnnm* $a = 11.737$ Å, $b = 14.550$ Å, $c = 3.451$ Å

Fig. 18 Electron microscope image of titanium selenide. "HREM" just means "high resolution electron microscope." Additional processing gives other images. Those are atoms. The microscope is operating at 300 kV, which gives 170 pm resolution (Thomas E. Weirich [Public domain], via Wikimedia Commons)

In *Weird Reality*, we also pointed out that *electrons* have a wavelength, which goes down as electron energy goes up. Using this fact and some very clever electric and magnetic lenses, people have gotten down to 100 pm this way, about half the size of an atom! In Fig. 18 you can see individual atoms! Figure 19 shows the range over which you can use this marvelous invention.

Remember, *there's nothing really there but forces—yet you can see 'em!*

Fiber Optics

I'm just going over the top of this one here. It's a whole other field, could be another chapter, can be a book.

An optical fiber has a high refractive index core inside a low index cladding. That way, most of the light stays trapped (Fig. 20). Fibers connect our civilization now (Fig. 21). What's special about a fiber? Remember from *EM Waves* that visible light is 400 THz or so. Because of limits in the boxes connected to the fiber, you can only put 1 THz of information on a fiber today! But, that's 62 million conversations, or 52,000 TV channels! You get my drift? The theoretical limit is about 25 times this over 10,000 km distance.

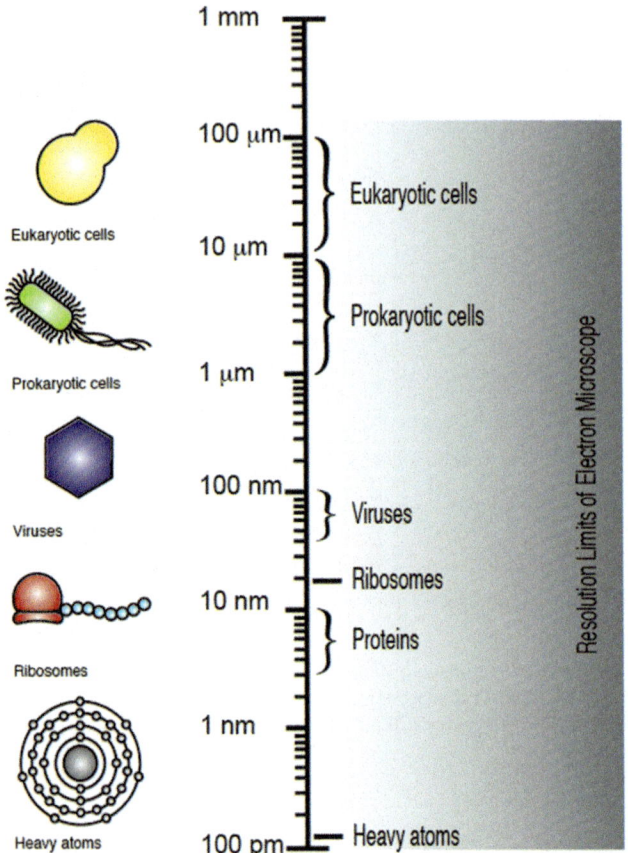

Fig. 19 Electron microscopes can see a lot of things (Jmgrants GNU free documentation license via Wikimedia Commons)

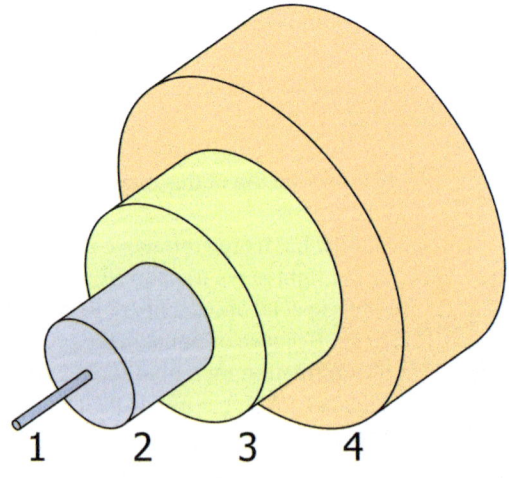

Fig. 20 Optical fiber. 1, 2, 3, and 4 are the core, cladding, buffer and jacket. The core is a little bigger than the wavelength of light (Bob Melish derivative from Benchill via Wikimedia Commons)

Fig. 21 Optical fiber bundle (Bigriz via Wikimedia Commons)

Of course, like with all concentration, this one increases vulnerability. Every so often in Santa Fe, some guy digs a hole where it says "don't dig here," and the whole city goes down from cutting one fiber. Take your pick—speed or robustness—that's the choice we make every day, and it's all toward speed.

Conclusion

We covered everything from selfies to telescopes, electron microscopes and 3D in this chapter! For related stuff, check out the *Lasers* chapter. There's so *much* more amazing stuff optics can do that I'd love to tell you about, but this chapter's too long already!

Now: lasers! The most amazing invention in physics or optics if you ask me, and important enough to have its own chapter.

Lasers

Theory

Have you wondered how lasers work? Maybe not, because you thought you'd never understand it. Well, here's where you're wrong on that!

Is anything magically special about laser light? Can it heal or cure in some way ordinary light cannot? Why can it do welding, cutting, drilling of teeth and blinding of human eyes—hard to do with a light bulb? Why can it shine to the Moon and back brightly enough that we can detect the flash on Earth? What is an excited state? Or stimulated emission for that matter? What did Einstein have to do with lasers back in 1917, a century ago and 40 years before anyone thought of lasers?

In this chapter, we'll answer some of these questions.

Einstein was here

F. Wicke

Spectra

Back in 1917, before anyone thought of a laser, Einstein theorized that a single excited atom can return to a lower energy state by emitting photons, a process he called spontaneous emission. He also understood atoms will only absorb photons of a certain wavelength. People knew that since the first high quality spectra (Fig. 1). Isn't that pretty? And you thought (unless you have looked into a spectrograph) that these were just colors. Each dark line in this Figure is because photons at that exact wavelength are absorbed by atoms in the sun, or atoms and molecules in our atmosphere, while the light is on its way from the sun to us. When a photon is absorbed, it makes electrons orbiting the nucleus of the atom go to a higher energy level. Absorption sets the stage for spontaneous emission. The electron really wants to not be excited and just get back home. But it can't do that

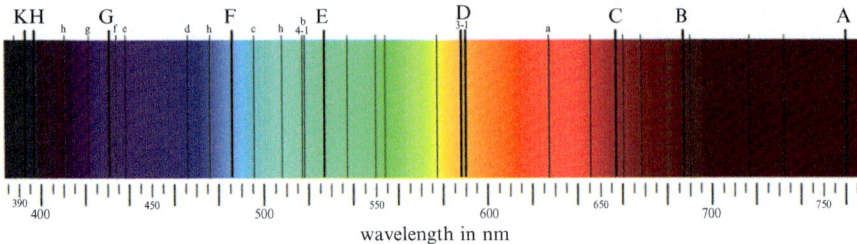

Fig. 1 Absorption lines in the sun's spectrum (Public Domain)

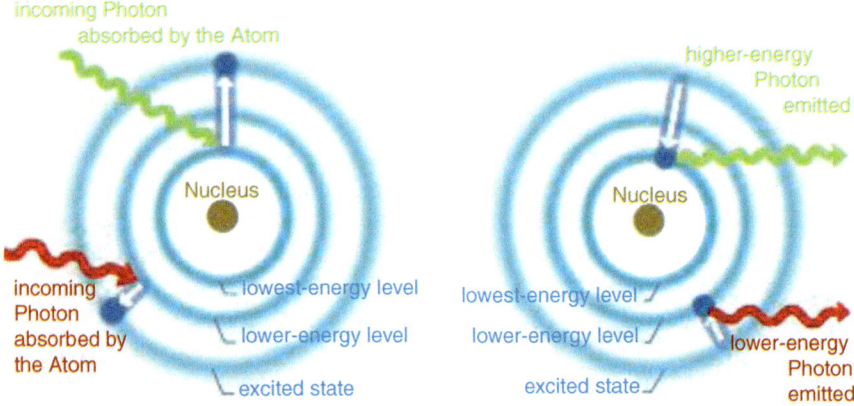

Fig. 2 Absorption and emission (F. Wicke)

instantly for various reasons. It may be that it can't go down instantly because that's a "forbidden transition" [Sounds exciting, no?]. Or it just takes a while for other reasons. When it does come down by itself, that's called fluorescence. If you still remember those, you know from the numbers on a watch dial glowing in the dark fluorescence can take hours.

Absorption

Oh. Did I say higher energy level (Fig. 2)? Sounds a little new-agey, huh? An atom's electrons (and therefore the whole atom) are in different energy levels according to their distance from the positive protons in the nucleus, and some other details of the orbit itself. Just as a manmade satellite has to fight against *gravity* to get into orbit, the electron has fought against the *electric* field of the nucleus, if it has risen to an excited state.

Quanta

There's only one difference, and it's an important one. As people like Planck and Bohr showed, thinking about those sharp lines in Fig. 1 spectrum, tiny things can't have just *any* energy, but just certain precise ones (quanta)! This is hard to understand at first, and it was for them. The energy that you or a satellite have is also *quantized*, but the number of quanta is so big that the difference from one to the next is too small to matter. In an atom, that's no longer so.

Photons also have a very precise energy, and it's a universal constant h called Planck's constant times the photon frequency ν. The bluer it is, the higher ν is, and vice versa [Check out Table 1 of *EM Waves*]. Frequency is just the speed of light divided by the wavelength λ. Absorption is mostly [never say only!] going to happen if that photon has precisely the energy needed to raise a particular electron in a particular atom from one orbit to the next and, to repeat, that is a precise energy. So: the frequency (color) of the photon that is absorbed by a particular atom is precise, too!

That's why those lines in Fig. 1 are narrow and black: the electrons in those atoms suck out all the light of a particular color. Each atom has its own fingerprint of "lines." So: if you know the atoms pretty well, when you see a group of absorption lines, you can say "Aha! That's sodium!" (Or beryllium, or iron 17 times ionized! It does get complicated!)

How can we use that? The "ChemCam" on the Mars Rover, developed by Roger Wiens of Los Alamos and 78 other folks from all over the world, shoots a laser at a rock, looks at the spectrum and tells what it is! Isn't that amazing?

So: back to our electron fighting its way up. It can fight in many ways: by absorbing a photon, by being hit by an electron beam from outside, by colliding with another excited atom, or by just getting very hot, as examples.

Chemistry

Electrons all have their homes in the atom. In hydrogen, it's pretty simple (one electron, one home). In a more complex atom like oxygen (Fig. 3), the natural state is two on the inside and six more in the next orbit up. For reasons you'd be bored with, the first orbit is full when it has two electrons, and the second orbit is full when it has eight electrons. But oxygen has only 8 plus charges in its nucleus, so it can only have 8 electrons, not its full 10, and it is unsatisfied all the time. To be satisfied, it would need 10 plus charges in the nucleus. It would then have a complete outer shell—but it would be neon, not oxygen (Fig. 4).

Because it's complete, neon doesn't normally react with anything chemically speaking, and it's called a noble gas.

Fig. 3 Oxygen atom
(F. Wicke)

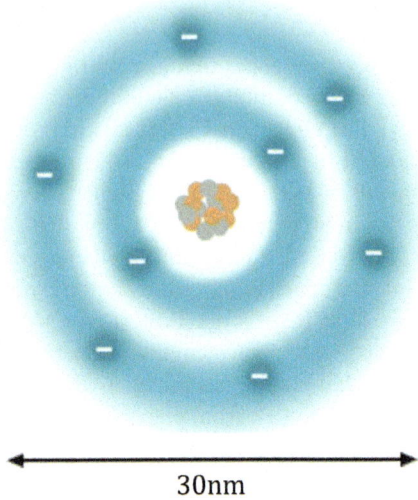

30nm

Fig. 4 Neon atom
(F. Wicke)

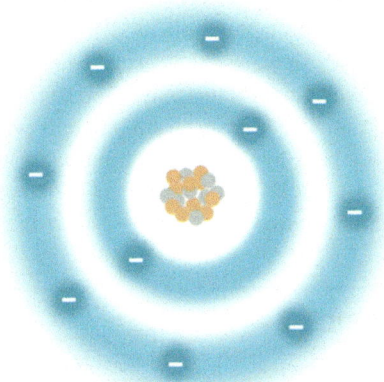

I can't help digressing for a moment to mention that we living things are so lucky that oxygen is like it is, hungry. It just wants to get two more electrons and the positive charges that would hold them, to be happy. How can it do that, short of a thermonuclear reaction? It cheats! If it can snare a couple of electrons belonging to an atom that burns (hydrogen for example), those atoms bring the plus charges with them, and the oxygen feels complete and understood. Figure 5 shows how this works with water. Can you imagine the world without water? Oxygen also combines

Fig. 5 Water (F. Wicke)

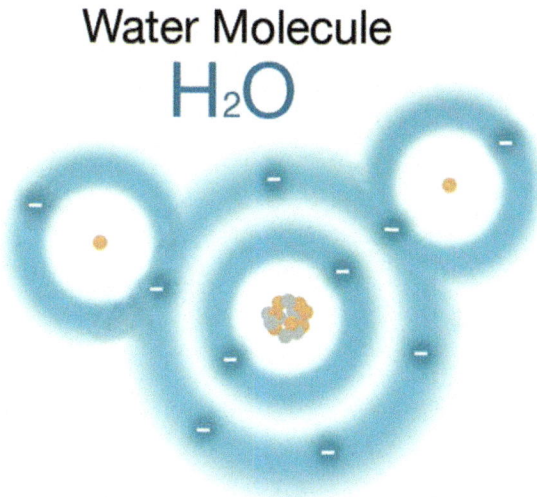

with lots of other stuff, like carbon. Then you get CO_2, warmth around campfires for millions of years, the Industrial Revolution—but also climate change when you burn too much too fast. Or you get rust when you leave iron out in the weather.

Emission

An excited electron can dump that extra energy randomly ("spontaneous" emission).

But it can't make the final transition and crash right into the nucleus. Why? We talk about that in *Weird Reality*. It's not obvious!

Einstein was here

F. Wicke

Or (an important "or") if a stray photon at the correct wavelength just happens along [incoming orange photon in Fig. 6], Einstein guessed in a genius moment that it can *stimulate* the atom to release its photon instantly. This does not destroy the incoming photon. After this, those two photons will travel in *the same direction* with the same frequency and phase as the original stray photon (Fig. 7). As I say in the *Weird Reality* chapter, an old girl friend thought "stimulated emission" a funny name for this process.

Another way to say it is that photons like to travel together, and in the same state. Doesn't that sound cozy? This is important for what a laser can do. This is genius—and there are a lot of things that come out of that preference!

"Same phase"? That just means they're both experiencing the same part of the rise and fall in the electric/magnetic wiggle that constitute *Electromagnetic Waves* at the same time.

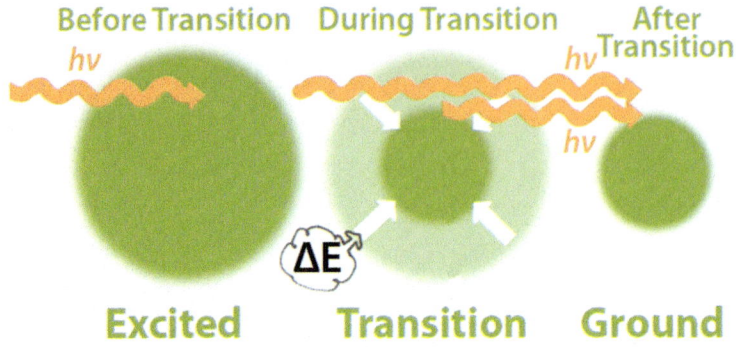

Fig. 6 Stimulated emission (F. Wicke)

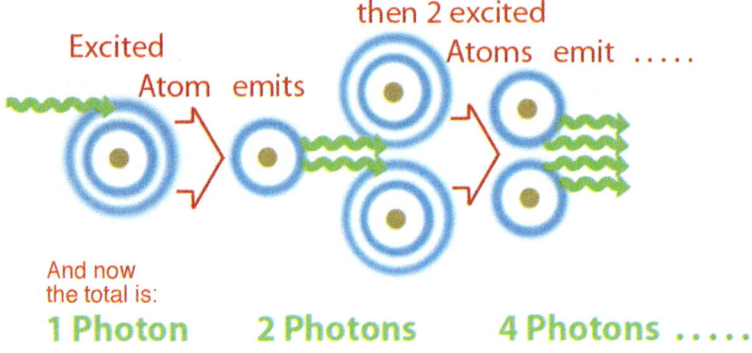

Fig. 7 A boxful of excited atoms is a laser! (F. Wicke)

Oh, ahh—where did the electrons go in Fig. 6? We are so *used to thinking* of them as tiny dotlike things whizzing around the nucleus. Now, we're going to wean you from that idea. I'm here to tell you it's a probability cloud and not a thing at all [see *Weird Reality*]. Not only that, like I suggested earlier, the distance from the nucleus is not the only unique property an electron has in an orbit. That electron wave can have different distinct ripples in its shape around the nucleus, as well as different "spins," and those differences put a limit on the number of electrons in an orbit, as well as slightly different energies.

Electrons don't like very close neighbors, but photons do! *The biggest division in the physical world is between these two personalities, the ones that like intimacy and those that don't.* These are bosons and fermions. We won't go into that deeper here.

The Laser Idea

Now, there are two photons! You can see from Fig. 6 that one photon becomes two. Two then become four, then eight ... (Fig. 7).

Does this remind you of Rani and her grains of rice? It should! [see *Exponentials*]. What happens in a boxful of excited atoms that have all been pumped up into a higher energy level and are all just waiting to be stimulated? Wow! At the speed of light, you've suddenly got a huge wave of photons of the same color moving in the same direction and sweeping up all the available energy in that boxful of atoms! From one photon to enough energy to burn something in millionths or billionths of a second! That's all there is to a laser!

Inversion

You may think some of your friends are inverted, right? Well atoms get that way too when for one reason or another there are more electrons in a higher state than in a lower state. This does not happen by accident—just getting hotter won't do it, because however hot it gets, there will always be more in the lower energy state than in the upper. That's just the way it is. But if I come along and pump the atom with an intense stream of high energy photons, or an electron beam, or a chemical reaction, I can make a "population inversion," where momentarily there are more electrons or molecules in the upper than in the lower state. You can understand that an inversion is necessary to make a laser. By the way, it's easier if your atom has an upper state where the electron gets "stuck," in a "metastable" state so that the transition back down is "forbidden," until it's stimulated. Not all lasers are atomic.

Start of the Laser Age

The question is, why did it take 40 years after Einstein for physicists and engineers to make a laser? Townes, Schawlow, Javan, Maiman ... in the late 1950s, just as Elvis was very hot, all these guys suddenly thought of it, and some did it, at the same time.

I'm not sure about the answer to my question, but I think it was because people couldn't think of a way to make an inversion.

All these guys thought about it, but in 1960, Ted Maiman at Hughes Research in Malibu went into the lab and did it! He built the first ruby laser, and it could burn the emulsion off a photograph in one loud pop. He used a pink ruby surrounded by a coiled flashlamp from G.E. (Fig. 8). The flashlamp made the inversion, and the ends of the rod itself acted as mirrors to bounce the light back and forth through enough distance for Rani and her grains of rice to go from one photon to maybe ten billion billion. In those days, the output was a series of spikes rather than a single giant pulse—but it was done! On the far wall Ted saw a pink spot.

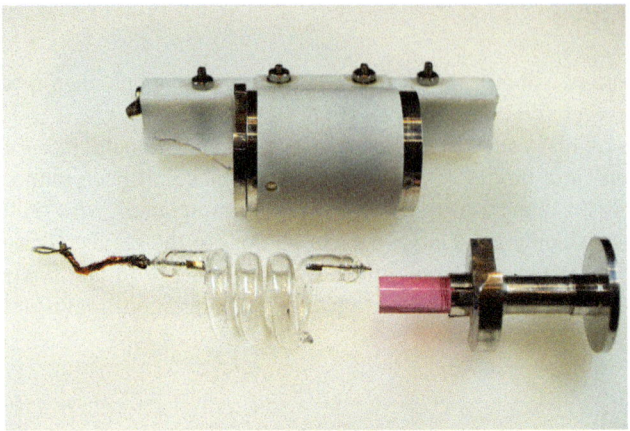

Fig. 8 The first laser (Guy Immega, Wikimedia commons)

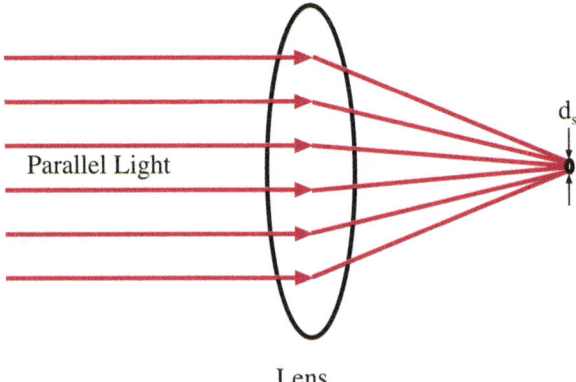

Fig. 9 Focusing bright light (C. Phipps)

Among Americans, theoretical physicist Charlie Townes got the Nobel for the laser. The Russians Basov and Prokhorov shared it. Not Maiman, who did it first, or Javan, who made the first gas laser. Go figure.

The Laser Zoo

From these humble but courageous beginnings, there are electrically driven gas lasers (CO_2, helium-neon), chemical lasers, solid crystal lasers—even the atmosphere of Mars has been accused of being a laser. Some of them are monstrous, like the 100TW NIF neodymium:glass laser (see later). Some are tiny glass fibers with lasing cores smaller than a hair. Some are little laser diodes that can make jets

The Laser Zoo

Table 1 The Laser Zoo

	Lasant	Host	Wavelength (nm)	Noted for
Solid crystal	Er^{3+}	Glass, YAG	1500, 2900	Many wavelengths: 2.9 μm is "eyesafe" (eye lens won't transmit it), 1.5 μm is for optical fiber communication.
	Cr^{3+}	Ruby	694	First laser. Few joules of energy pulsed
	Ti^{3+}	Sapphire	700–1000	Huge bandwidth makes it suitable for ultrashort pulses
	Nd^{3+}	Glass, YAG	1065	Can be CW or pulsed. Mainly noted for giant energies.
	Yb^{3+}	Glass, YAG	1065	Same apps as Nd:glass
Semi-conductor	GaAs, InGaAs	–	780–905	Semiconductor laser diode. Tiny, tens of mJ pulsed, tens of W CW. Main use in communication and laser pumps.
Gas	H_2O	–	0.028, 28,000, 220,000	Yes, water (vapor) can lase!
	CO_2	–	9200–11,400	W to tens of kW CW. Used for welding and cutting.
	OI	–	1315	CW, oxygen-iodine chemical laser, MW power.
	HeNe	–	633	Helium-neon laser, Javan's discovery, tens of mW CW. Good for alignment, surveying.
	Cu	–	578	Copper vapor does the lasing.
	Ar	–	458, 476, ... 502	Tens of W of CW power
	Kr	–	416, 458, ... 799	Tens of W of CW power
	XeF	–	351	"Excimer." Hundreds of J pulsed.
	XeCl	–	308	"Excimer." Hundreds of J pulsed.
	KrF	–	248	"Excimer." kJ pulsed.
	ArF	–	193	"Excimer." Tens of J pulsed.
	F_2	–	157	"Excimer." Tens of J pulsed.

to push spacecraft. Some are "excimers," momentary compounds like XeF or KrF that ought not to exist. Table 1 lists some laser types you might be interested in.

Terminology: YAG: yttrium-aluminum garnet, GaAs: gallium arsenide, InGaAs: indium gallium arsenide, Er: erbium, Yb: ytterbium, Nd: neodymium. Did you even know those last ones were elements? Excimers include XeF: xenon fluoride, XeCl: xenon chloride, KrF: krypton fluoride and ArF, argon fluoride. Lasant means what lases, host is the crystal in which it is embedded for crystal lasers. Then, there are *molecular* lasers. In CO_2 for example, energy levels come from different molecular stretches and bending, which I won't drag you through here (Fig. 10).

There are also x ray lasers from 3.5 to 45 nm. Too much detail? Here's a visual:

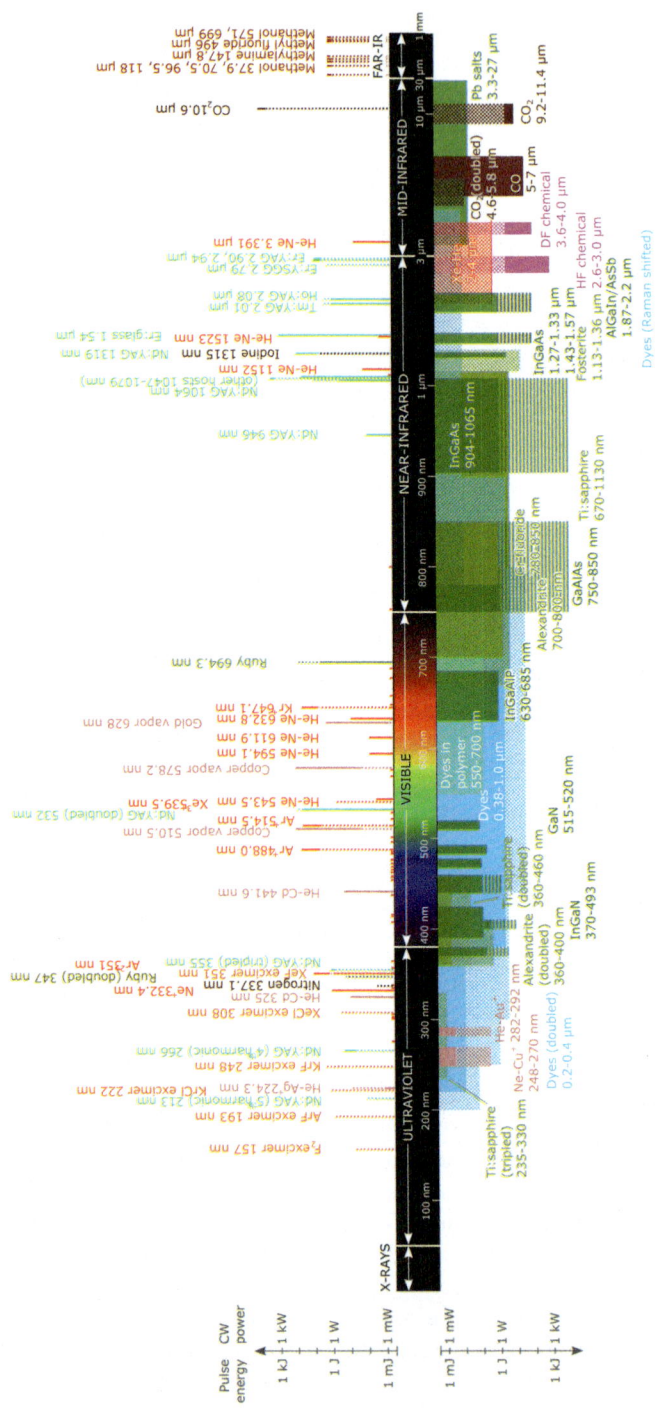

Fig. 10 The laser spectrum (Danh, Wikimedia Commons)

What's Special About Laser Light?

1. It's very *bright* because the mirrors on each end of a laser oscillator make the light go in one direction, the same as that first stimulating photon.
2. It is usually just *one color* (except when it's not, as in the Ti:sapphire laser we talk about in the next section.

What do these features let us do? (Fig. 9)

Because the light rays are parallel, a good lens can focus them to a spot with diameter as small as a wavelength of light! If that is 1 μm, and we have just 1 W of power, the intensity at the focus is 100 MW/cm^2! You can use this feature to *cut*, *weld*, or do *laser fusion*, where you are trying to create temperatures of hundreds of millions of degrees, like in the very center of the Sun! You can make welds *inside* a tube, or very fine and intricate patterns on solid materials, like integrated circuits.

You can't focus an ordinary light source—even the Sun—like that. Ordinary light comes from a hot filament or an electric arc, in any case something with a hot surface at temperature T putting out rays in all directions. Thermodynamics says there is no way I can focus that to produce a temperature higher than T! If I could, then heat would go back from my focus to its source making it hotter, and that would be ridiculous.

So: *a laser is brighter than the sun*! Think about that. You can hurt your eyes. You can make a *plasma* [what happens when you make something hot enough to become the fourth state of matter, just ions and electrons]. You can use a laser-produced plasma for *space debris removal*. We'll get into that later.

The narrow linewidth can let us pick just one chemical reaction and make it go. As an example, for cancer therapy, we can inject something that binds to the cancer and also kills tissue when we put light of a certain wavelength on it (*photodynamic therapy*). Any wavelength won't work. Just the one that starts that reaction.

Ultrashort Pulses

You can skip this section if it looks too hairy.

Shorter, and shorter, and shorter pulses and more and more energy and colors (wavelengths) ... this is the history of improvements in lasers. Maiman's laser pulse was milliseconds in duration. First Q-switching, then modelocking took that to microseconds and then nanoseconds. To make an ultrashort pulse, you need an ultra broad bandwidth laser like titanium-sapphire. That's because the uncertainty principle says it's so [see the section on uncertainty in *Weird Reality*].

Up to now, I've been talking about precise, narrowband transitions in single atoms. When the laser is an impurity ion in a crystal, it's more complicated.

The ruby laser is aluminum oxide crystal with a few percent chromium atoms that get three times ionized just by being in the crystal. The Cr^{3+} ion is the lasant.

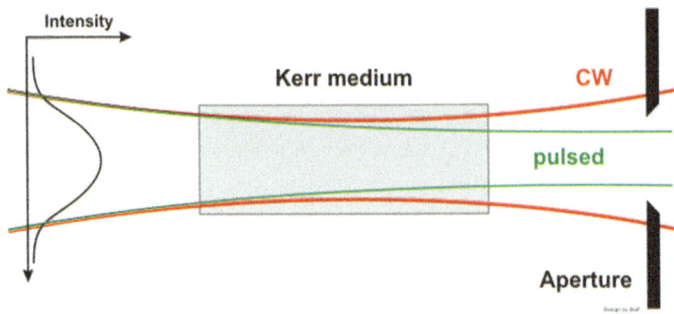

Fig. 11 Kerr lens modelocking (Creative Commons license)

Titanium sapphire is an aluminum oxide crystal with Ti^{3+} ions replacing aluminum atoms, instead of Cr^{3+}. In the crystal, there are powerful electric fields that hold the ion in place. So powerful that they can change the laser frequency depending on the distance of the impurity ion from the nearest oxygen atoms in the crystal structure—and that can change because of crystal vibrations. Because of this, Ti:sapphire can lase from 700 to 1000 nm wavelength, let's say 850±25 %! In a very short pulse, this beam will have more than all the colors of the rainbow—if you could see out to 1000 nm in the red. That center wavelength corresponds to 350 THz and the 50 % bandwidth corresponds to a pulse only 5 fs long! *It takes that much bandwidth to make a pulse that short.* In such a short time, light only travels 0.3 µm.

But how do you get that short pulse? One way is a brilliant invention called the Kerr lens modelocker. Laser beams are stronger in the center and weaker at the edges (Fig. 11). In the Figure, "CW" just means low intensity, and "Kerr medium" means something which causes a very intense beam to focus itself, because the refractive index increases with intensity. That can happen in femtoseconds, allowing a pulse that short to get out of the aperture. The actuality is a little more complex with multiple mirrors on both ends.

This is a fs oscillator, and it will have a pretty small energy.

Now how do we get a lot of energy in a short pulse? With an amplifier, duhh! But how do I do that without more self-focusing? Gérard Mourou invented chirped-pulse amplification in the 1980s to solve this problem, and that has given us tens of joule amplified beams down to a few femtoseconds width, and that means petawatts! The secret here is just to reduce the intensity by spreading the beam in time while you amplify it—so it won't self-focus in the amplifier—then recompress it. How do you do that? We said it has all the colors of the rainbow, right? Have you seen a diffraction grating? You can arrange a pair of them in such a way that red colors get through quicker than blue ones (Fig. 12). What comes out is a million-times-longer pulse that is red at the front and blue at the back, with much lower peak intensity. That's chirping. Now you amplify the beam. Then you put it through a pair of gratings arranged so the reverse is true and you have your powerful femtosecond pulse! I admit it's complicated.

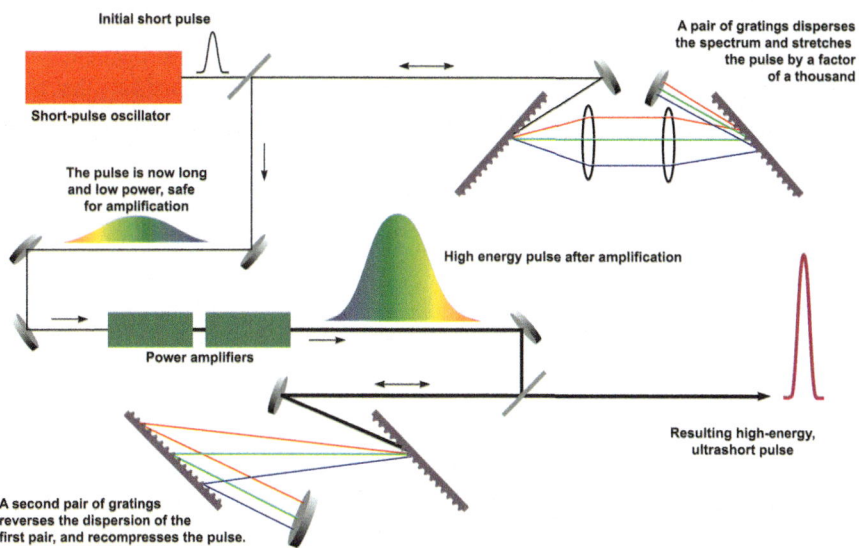

Fig. 12 Chirped pulse amplification (US Department of Energy—Public Domain)

Fig. 13 A bat (Oren Peles, Wikimedia Commons)

Digression: Bats and Chirped-Pulse Acoustic Radar

This is the same thing bats do (Fig. 13). They use acoustic radar, and they can locate little mosquitoes and so on 1 mm in size. But at the speed of sound in air, that's a few microseconds, and the bat can't make so short a click. What does he do? He chirps and then reformats the chirped echo from a mosquito using a dechirping circuit in his ears and brain so that he registers a click. This is all at frequencies we can't hear, so you aren't aware of it. But in South Africa, I've heard bats that chirp in the audible range.

Petawatt Pulses, Attosecond Lasers

Several labs around the world have now made beams with 30 J in 30 fs, which is 1 PW! If you focus that to a square wavelength of light, you've got 1E23 W/cm^2, a truly astronomical number. You can circularly polarize that beam, split it and recombine it in a cloud of neon or xenon gas in such a way that the two beamlets only add up to a big electric field during one optical cycle, about 1.3 fs.

The electric field of the light itself can be 10 GV/cm (see *Electromagnetic Waves*). This is 100 V across the size of an atom, and it only takes 10 V or so to ionize an atom. So, when that electric field hits the poor neon atom, it can strip an electron right off it, after which it comes crashing back down on the nucleus and—¡bang! you have an attosecond pulse! That's because acceleration is what makes a moving charge radiate, and hitting that nucleus is an extremely abrupt stop.

Boy, this sounds complicated. So we're going to leave it at this. Without getting into it deeper, 50 as is how short pulses are now!

So OK what can I do with a 50as pulse? In such a short time, I can catch chemical reactions with their pants down, so to speak, literally take stop motion photos of every stage of a reaction, at 100 million frames a second, as someone recently did.

Ohh—and a pulse *that* short must have an even broader bandwidth to exist, compared to Ti:sapphire, so it will be an x ray pulse!

Applications

Now that we've talked about some of the different types of lasers, let's go to specific uses. Just like for our laser zoo, the story will go from low to high power, continuous to short pulses.

Photodynamic Therapy (PDT)

Figure 14 shows PDT being used. The patient has been injected with a light-sensitive drug that finds and binds to cancer cells. Then, the doctor uses an optical fiber to carry light of just the right wavelength from the laser to the cancer. What happens then is that the drug becomes a poison that destroys the tumor without damaging normal cells very much. Isn't that amazing? Doesn't work for all cancers, of course, brain cancer in particular, because the "blood–brain barrier" keeps the drug from getting from the blood to a brain tumor.

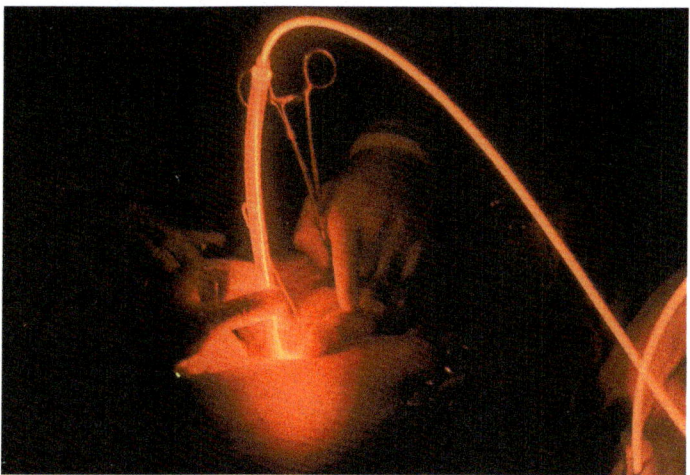

Fig. 14 Photodynamic therapy (National Cancer Institute, public domain)

Laser Engraving, Welding and Cutting

Because of Fig. 9, focused light from a few-kW laser can get things very hot, or cut very finely.

Laser engraving is faster. more accurate and more intricate than any other way. You can engrave wood, glass and stainless steel.

In welding, if you put an inert gas like helium over your work, the laser isn't adding any impurities as can happen with a gas or electric welder, and air can't oxidize the hot surface. You can weld aluminum and steel together! These days, fibers can carry tens of kW of light to a workpiece and it's easy to do this fast under robot control. The shell of your car is probably welded together this way.

But where's the laser light in Fig. 15? It's invisible, has 12 kW of power, and it's coming directly down on the workpiece from above, not through one of those nozzles. The big nozzle is bringing inert gas to the weld spot. The small nozzle blows away fumes. On the surface, you see a laser-produced *plasma*. Bear in mind that invisible beams are more usual than not—these days the most common laser welding and cutting beam will be 1.06 μm, from a diode-laser-pumped glass laser, and that's about 50 % too far into the red for you to see. That's why they're dangerous.

3D Metal Printing

Sometimes, laser welding is the only way to do it. For example, welding the *inside* of a tube. And you've all heard about *3D printing*. This doesn't use a laser, but just squirts plastic onto a mandrel according to a computer design. But if you want to make a drill bit with amazing properties, you can squirt a metal powder onto that mandrel and melt it with a laser, *making 3D metal parts* that can't be made any other way.

Fig. 15 Laser welding (Krorc, Creative Commons license)

Cornea Surgery

If you need your cornea reshaped, microsurgery using an excimer laser is the way to do it. At those short wavelengths, the laser light actually lifts off (ablates) the cells on the *inside* of the cornea, so you get a new cornea shape without burning.

Laser Gyros

Next time you fly across the Atlantic, think about the fact that a laser gyro is telling your pilot where you are. Inertial guidance is one of the most ingenious, magical things mankind has invented: a device that stays pointed at the same spot in absolute space, even as the Earth turns underneath it and, with the help of computers, always knows exactly where it is. It's GPS without the GPS! How would you use it? You know the exact locations on each end of the trip in three-dimensional absolute space. You select the destination, program it in and tell it to go.

Another reason you would want such a thing? You're going from the Earth to the moon, and there are no GPS stations. Also, unfortunately, since the days of the V2, ICBM's have needed gyros to know where they're going without external references that could be jammed.

How does it work? There are two paths the light can go (Fig. 16): clockwise or counterclockwise around this loop. I never said a laser had to lase in one direction. Now, imagine the table underneath it is rotating clockwise. Each time a clockwise photon goes around, it takes a tiny bit longer than the counterclockwise photon to get to the same spot. This is because the goalpost (the detector behind the partial mirror) moved during the time the photon was going around!

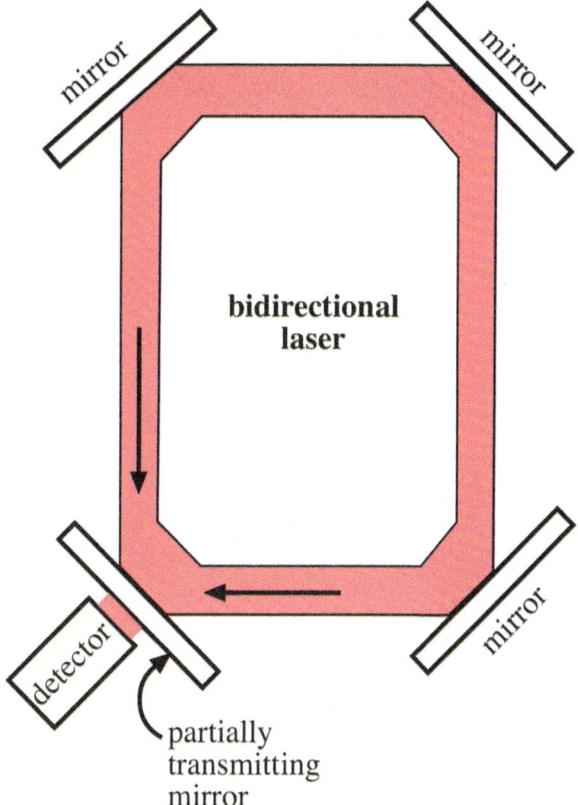

Fig. 16 Laser Gyro (C. Phipps)

The difference is a fraction of a wavelength of light, but the detector can see that difference as the waves add and subtract. Believe it or not, these things can now detect 0.01° per hour rotation, and have a 60,000 h lifetime.

Knowing the angles so well and the distance it traveled in three dimensional space (there are three mutually perpendicular gyros), the computer can compute where you are in absolute space, then subtract the rotation of the Earth in the meantime and say how far you are from Paris. Isn't that amazing?

Here is what one actually looks like (Fig. 17).

Before this genius invention, people used very, very accurately machined spinning wheels. The mechanical gyros made for Apollo were designed and built at MIT's Instrumentation Labs. I was privileged to be present May 25, 1961, when the I-Lab public address system broadcast President Kennedy's announcement that we were going to the Moon. Everyone there knew how important that would be for the nation and for us. This thing was not perfect. It drifted about 0.05° per hour (five times worse than a laser gyro!) and had to be reset periodically by the astronauts

Fig. 17 Laser Gyro in operation (Nockson, Wikimedia Commons)

sighting on a star. They chose Canopus, one of the brightest stars in the Southern Hemisphere. It didn't matter that Canopus is hard to see in the USA; it was going to be easy to find between the Earth and the moon.

Laser Produced Plasma

Light *intensity* does stuff. You can shine a flashlight on a piece of paper at 1 W/cm^2 all day and never see a plasma or anything else. Use a good lens to focus the sun on a piece of paper (1 kW/cm^2) and you'll make it flash into flame. Focus a 25 kW Nd:glass beam onto steel with a 2 mm focal spot (1 MW/cm^2) and you'll weld it. These are all continuous (not pulsed) light sources.

If you have a pulsed laser, you can put just 100 mJ of energy in a 1 ns pulse on that same 2 mm spot and you will get a plasma, the fourth state of matter as temperature increases, after solid, liquid and gas. If you make a plasma jet on a surface with a pulsed laser, you get pressure, and that pressure pushes back on the surface. If the surface is a piece of space debris (Fig. 18), you can push it around. If you do it right, you can slow it down just enough that it will re-enter the atmosphere and burn up. If these pulsed beams are brought in from every direction around a sphere of DT (deuterium-tritium), then you compress it and make a miniature hydrogen bomb. We'll visit all of this now.

Orbital Debris Removal

One of my favorite projects has been laser space debris removal. The laser beam has kW of power divided into a string of very high power short pulses that act like a machine gun pointed at the debris (Fig. 18). You can also put the laser in orbit, which now seems a lot easier way to do it than having the laser on the ground because you can work at closer distance without the air.

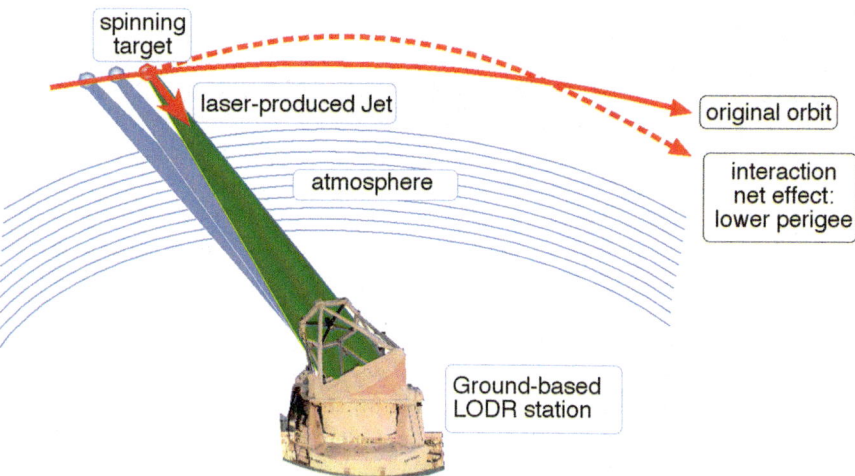

Fig. 18 Laser Orbital Debris Removal (C. Phipps, Photonic Associates, LLC, by permission)

Fig. 19 The Laser Plasma Thruster (C. Phipps, Photonic Associates, LLC, by permission)

Laser Space Propulsion

You can also use pulsed lasers to make a plasma jet on an object and drive it through space. Here is an example that uses semiconductor laser diodes shining on a piece of rolling tape like 8 mm movie film to make little jets. You can see the holes where the jets come out in Fig. 19. It weighs 1/2 kg, and makes just 10 mN using six diodes, 20 W of power and 30 m of fuel tape.

Pyramids of Cheops

I hate to put it this way, and I mean no disrespect to the folks who have designed and built these facilities, but three laser applications are of a size that would have made a Pharaoh proud. The first of these is the National Ignition Facility in Livermore, CA, east of San Francisco and north of San Jose.

NIF

You want to see a really big laser? Figure 20 shows the very largest one in the world. A ten story building as wide as three football fields and 120 m long contains a huge Nd:glass laser with 192 beams and a vacuum target chamber 10 m in diameter. This is truly the Cheops Pyramid of laser science. Each of these is then frequency-tripled into blue light. The result is 192 blue beams with total energy of 1.9 MJ in 20 ns. That's 100 TW! Every one of these go through different windows in the target chamber (Fig. 21) into the tiny container in front of the guy's eye in Fig. 22, arriving at the same time within 30 ps, if you can imagine that. Right at the entrance to the gold cylinder, you get about 5 PW/cm^2.

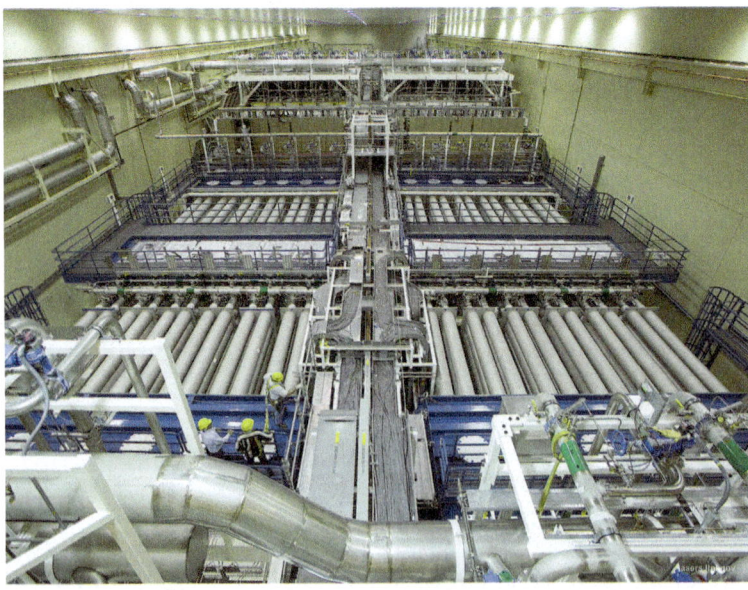

Fig. 20 *One* of the two laser bays at NIF, the National Ignition Facility (US Department of Energy—Public Domain)

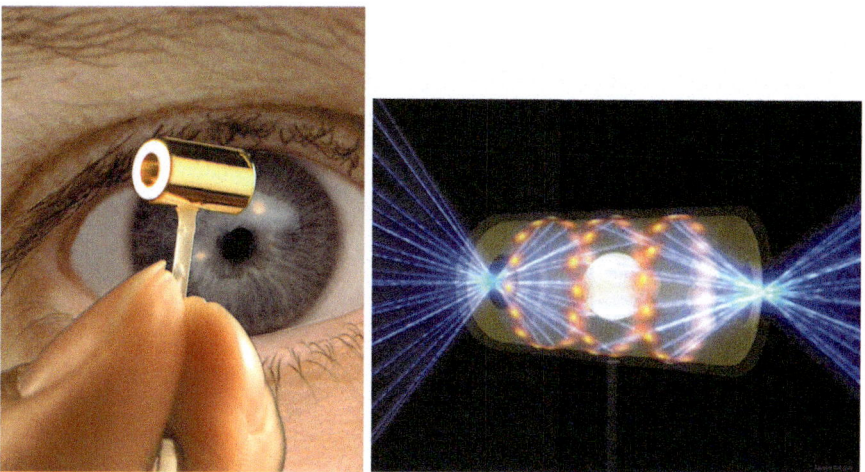

Fig. 21 NIF target container (US Department of Energy—Public Domain)

Fig. 22 NIF target chamber being delivered (US Department of Energy—Public Domain)

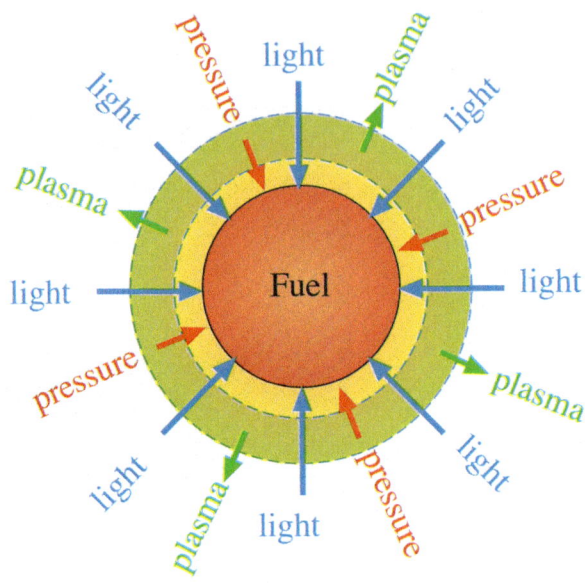

Fig. 23 Spherical compression from laser-produced plasma (C. Phipps)

Laser Fusion Dream

Now here's the point: The beams then spread out in the target container, heat the cylinder to x ray temperatures, and the heat from *that* makes a plasma jet all around that little sphere full of hydrogen in the center, compressing it to 100 times the density of lead, at 100 million degrees K and 100 billion atmospheres of pressure! (Fig. 23) Can you imagine that? It's like the center of the Sun. The hydrogen is actually deuterium and tritium, two heavy forms of hydrogen. It's called the National Ignition Facility because at those temperatures and pressures, the little target *goes off*—in a thermonuclear reaction like a miniature hydrogen bomb. You might get 20, or even 100 MJ out.

Crazy complicated as this all is, that explosion should make a *lot* more energy than came in. And because you can find the deuterium in sea water, the dream is making all the power we need for a billion years, literally.

Ah well. It's a wonderful dream, it's been 40 years of making larger and larger lasers at great expense, and some very good guys haven't made it happen yet. We'll see what happens.

Making Integrated Circuits

You may have read about this (Fig. 24). How does this involve lasers? Let me tell you a story. Have you heard about:

Fig. 24 An Intel 8742 12 MHz 8-bit CPU chip (10-year-old technology!) (Ioan Sameli, Wikimedia Commons)

Moore's Law?

Way back in 1965, Gordon Moore, cofounder of Intel, wrote a paper in which he guessed that the number of transistors in an integrated circuit doubles about every 2 years (Fig. 25).

Back in the time of the dinosaurs (1962), I visited what was then the Norden Corporation Research labs in Hartford. There, looking through a binocular microscope, I saw what looked like New York City with streets of gold connecting little square plazas. This was one of the first attempts to put a bunch of transistors and gates (switches) on the same semiconductor chip. Maybe it was 8×8—I don't remember, but incredibly primitive compared to the two billion transistors in the central processing unit (CPU) of your iPhone (Fig. 25), and 1962 was way before the data in the Figure.

In order to get that many transistors onto a chip of semiconductor that is only 2 cm on a side, guess what has to happen? Each transistor has to be scaled down to *400 nm* on a side, so the size of features you can draw on a chip has to go down to 50 nm! (Fig. 26) Yes, I said 50 nm. Not 50 mm, not 50 µm, but 50 nm! In fact, it is now possible (2014) to get 14 nm feature sizes, and 5 nm is predicted by 2020. At each point along this route, people have wrung their hands and said Moore's Law was finished, wondering how they could ever get smaller. But you can see it's not done yet. Still, the limit of what Nature allows in two dimensions is close at hand—the distance between atoms in silicon is about 1/2 nm.

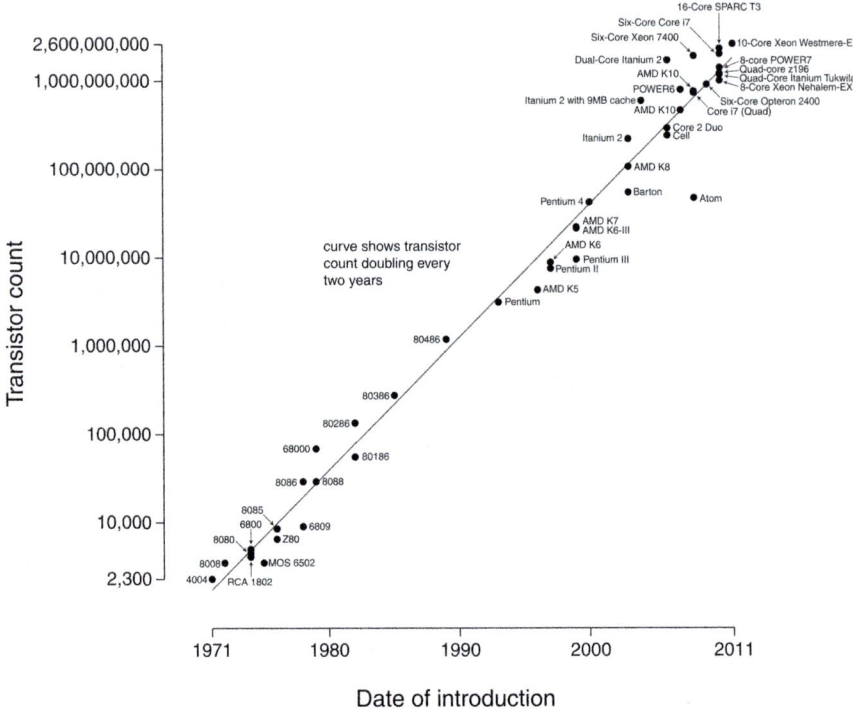

Fig. 25 Moore's Law (Ioan Sameli, Wikimedia Commons)

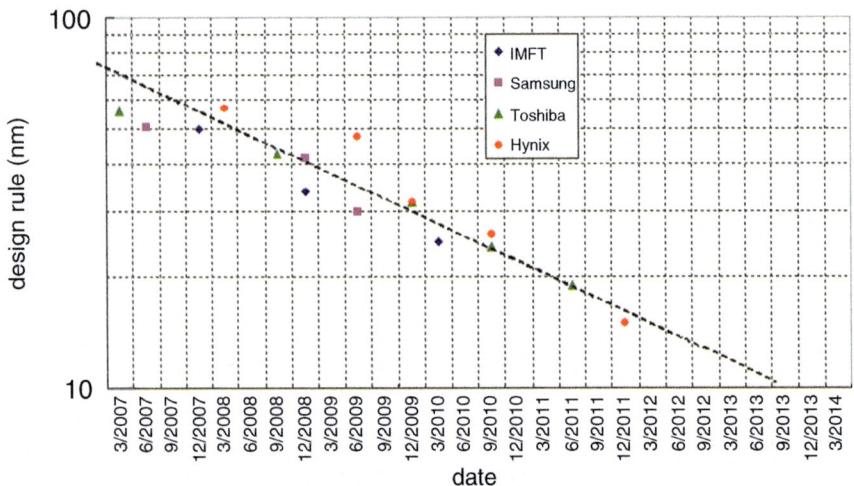

Fig. 26 Feature sizes (Guiding Light at English Wikipedia, Wikimedia Commons)

Chips are made hundreds at a time on a 30 cm diameter silicon wafer in vacuum, covered with a glass layer and a "photoresist." A huge complex pattern is projected reduced 5:1 onto that, developed, and the places where the photoresist is not are etched way, leaving gaps of pure silicon where the impurities that make this thing part of a transistor can be shot into the wafer. You do this several times to make an IC wafer.

Now, look again at Fig. 9! How do you do that projection? How do you use light with normal μm wavelengths, to make tens of nm patterns?? You don't. You can use electron beams. Or you can make a 13.5 nm wavelength light source, and that's where lasers come in. This is extreme ultraviolet. A CO_2 laser makes a laser produced plasma, of the sort we talked about earlier, in xenon gas or tin vapor. For Xe, the CO_2 beam is focused in the gas with enough intensity to heat it to about 450,000 K, which strips off ten electrons, leaving ten-times-ionized Xe^{10+} ions that put out 13.5 nm light! For tin vapor, the ions can be up to Sn^{13+}!

Please don't go to sleep on me here! This is how hard people struggle to make your iPhone!

OK. So we've got the EUV radiation, and we can even double-expose to get somewhat smaller features, maybe down to 10 nm. Oh, by the way, you have to use mirrors to focus this light—lenses are opaque. And, you have to have kW of CO_2 light to make the tens of W of EUV you need to have a reasonable wafer production rate.

Of course, if these technology advances are happening every 2 years, guess what? You need a whole new plant every 2 years to make the wafers.

Modern Fabs

Now for my second Cheops Pyramid analogy: in 2014, a new semiconductor fabrication plant (in Taiwan) costs about 9B$ to build. This gives you a clean room with 100,000 m² floor area. That's a 300 m × 300 m clean room, about nine city blocks if it were all on one floor. The buildings have 400,000 m² of floor space on a 20 ha site (Fig. 27). Get the picture? All for your iPhone! And you, engineer dreamer, better be darn certain your new technology works.

LIGO

The third Pyramid is LIGO, one of the most technically amazing laser applications. LIGO stands for *Laser Interferometry Gravitational Wave Observatory*. Do you remember *Weird Reality*, where I said there must be gravity waves but nobody has seen one yet, directly? This is the tool that will do it. At $600M, it's the largest single investment our National Science Foundation has made. If a wave did come through, it would probably shorten one arm of a Michelson interferometer (Fig. 5, *Electromagnetic waves*) more than the other. The arms of this one are 4 km long!

It's actually two observatories, one in Louisiana and one in Washington State, with an effective "telescope diameter" equal to the distance between these sites.

Fig. 27 A modern "Fab." (Peelden, Wikimedia Commons)

That way, if they ever do detect one, they can at least tell which direction it came from, and maybe how far it was. So far they haven't seen anything. The one in Louisiana is just being fixed up to be ten times more sensitive, so it can detect a length shift of a tenth of an attometer! Yes, that's right. A billionth of the size of one atom. To do this, everything is in a vacuum and mirrors hang on pendula which hang on pendula of glass fibers four pendula long. The 1064 nm laser will be 750 MW, because at low power, even the photons make noise! The rumble of the ocean on a distant shore is an important interfering noise. What they're looking for is something singing between 10 Hz and 10 kHz. That would come from a pair of neutron stars spiraling into each other in their death throes. They will be able to see this happening anywhere within *half a billion light years from Earth*. The event is so rare that they need to look into a sphere of space with this radius in order to see maybe ten events per year.

Now, come on! That's big. About 4 % of the radius of the Universe.

Conclusions

Lasers are, at least to me, the most fascinating invention of the twentieth century, and an amazing extrapolation from high powered theory to real devices of all sorts. They can do surgery, communications, engraving, laser fusion, laser space propulsion and iPhone CPU fabrication, just to name a few applications. All from a 100-year-old theoretical speculation of Einstein's.

Nucleonics

In *Weird Reality*, and *Lasers*, we talked about atoms, and chemical reactions in which the electrons orbiting atoms help them combine into compounds, or absorb energy and make *coherent photons* in a laser. We ignored the nucleus, which was always a tiny, positively charged clump in the middle.

Energetically speaking, the most that could happen out there in the atom's frontier amounted to a few electron-volts (eV) of energy exchange. As we zoom into the nucleus, we're talking about reactions that can involve millions (MeV) or even hundreds of millions of eV, because of the tremendous forces that hold the nucleus together.

Here, we'll talk about just two kinds of nuclear reactions: fission, in which a really heavy nucleus like U235 is hit by a neutron splits into the nuclei of two other different atoms, or fusion, in which two very light nuclei like hydrogen or helium are forced together to make a heavier element.

This really *is transmutation*, changing one element into another, the dream of sorcerers and alchemists for thousands of years, and actually accomplished in the last century!

Both fission and fusion make tens to hundreds of MeV of energy per atom. Why? As atomic mass goes up, the binding energy per nucleon at first goes up very steeply. Then it tops out at Iron, which has more than eight times the binding energy of a hydrogen nucleus, and drops about 15 % as we get to Uranium (Fig. 1).

Stay with me! You *can* understand this! First, some definitions:

Nucleon? A generic term for either a neutron or a proton. *Atomic mass?* One atomic mass unit (amu) for each nucleon. An amu is 1.66 yg (see *Numbers!*). Most common versions of lighter elements have the same number of protons and neutrons, for example, carbon with six protons and six neutrons with atomic weight 12 amu. U235 is 235 because the nucleus contains 92 protons and 143 neutrons for a total of 235 nucleons. U236 still has 92 protons, and a matching 92 electrons which is what makes it uranium, but 144 neutrons. U238 has 146 neutrons. They're all *isotopes* of uranium with mostly the same chemical properties set by the electrons, but different atomic weights.

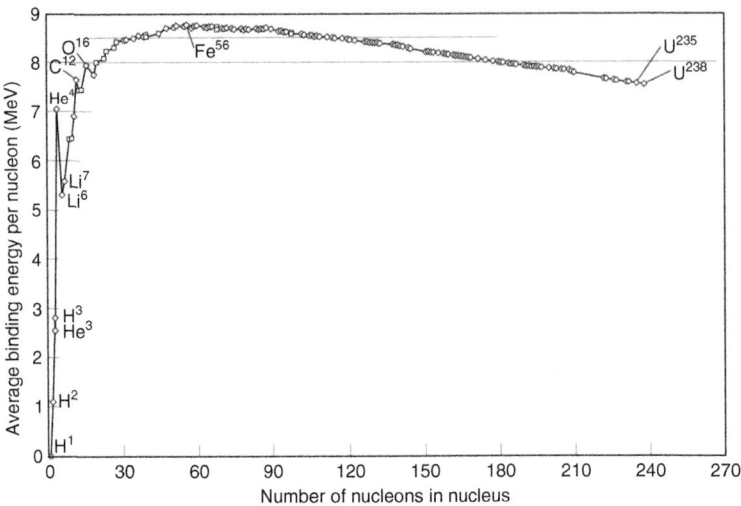

Fig. 1 Binding energy vs. atomic mass (Fastfission. Public Domain via Wikimedia Commons)

Now: *binding energy*. If something is more tightly bound, it has given up relatively more energy. I know it's counterintuitive, but … you climate folks will understand this—when a lot of ice forms in the ocean, a lot of energy has to be released so the air is warmer! You know that. For everyone else: think of steam in a kettle cooling into water and then becoming rigid as an ice cube. Those molecules are now as tightly bound as they can get. What happened? Just the reverse of boiling—energy was taken out, by a refrigerator of some sort. The back of the fridge is hot! As you go from nucleus to nucleus up the Fig. 1 curve, you have to emit energy! Those light atoms want to fuse, because that's a lower energy state! Same way, as you go down in mass toward the center of Fig. 1 from uranium, you also emit energy. U235 is a big, overweight blob, just waiting for an excuse to split!

Fission

When U235 splits, it becomes two atoms. One is barium, Ba141 and the other krypton, Kr92. Why those particular atoms? The answer is not interesting. The total of those two is only 233 amu, not the original 235, because two neutrons (each 1 amu) go flying away. The krypton is very near the top of the curve in Fig. 1, where atoms like to be. Each Kr92 is a bit more than 1 MeV higher per nucleon than it was as part of uranium. For 92 nucleons, that's 100 MeV in round terms. For the Ba141, it's less, maybe 0.75 MeV per nucleon but there are 141 of them. Add it up and you've got 200 MeV per uranium atom disintegration. That's a lot of energy!

There's another point here, which is why people thought of bombs: a chain reaction is possible (Fig. 2). Each U235 fission starts with one neutron and makes 2.6 neutrons on average. Those can fission more than twice as many atoms in the next

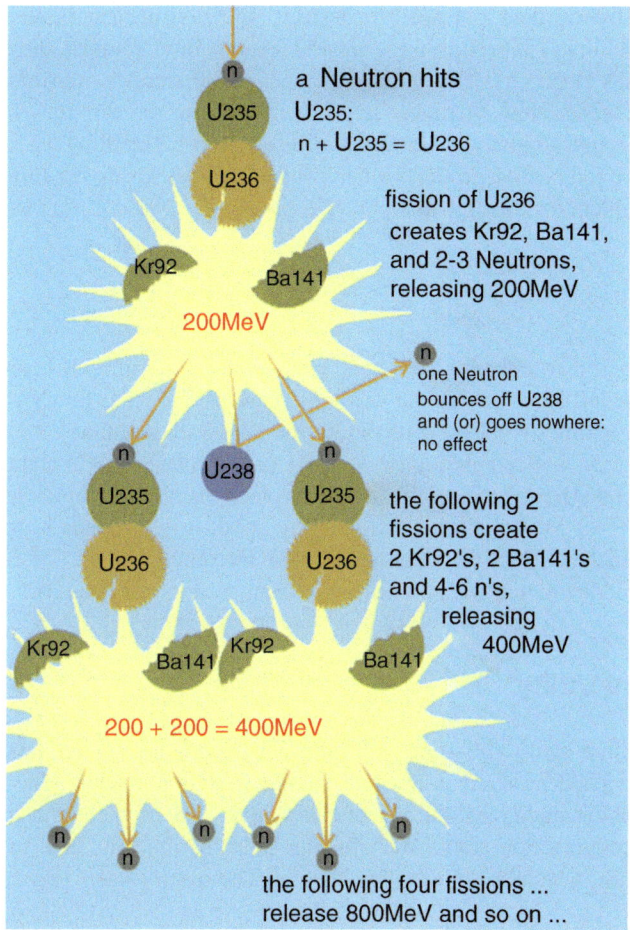

Fig. 2 Fission chain reaction (F. Wicke)

generation. This is a growing *Exponential*, and, for fast neutrons, it's going at a good fraction of the speed of light, just like the growing pulse in a *Laser*.

That was the exciting news from Otto Hahn in Germany in 1938, news he sent to Jewish scientist Lise Meitner in Sweden where she had just fled: uranium fission had been achieved. The cat was out of the bag. Einstein wrote Roosevelt and the Manhattan project was begun by 1941, resulting in a bomb 4 years later.

When a kilogram of uranium fissions, it releases a million times more energy than you'd get burning a kilogram of hydrogen. The energy comes out in neutrons and gamma rays. The sum of the masses of the products is 1/5 proton mass less than the mass of the U235 that started it. That difference *is* the 200 MeV of energy, by $E=mc^2$. Uranium has just converted 0.1 % of its mass to energy! Not much, compared to Fig. 7 of *Weird Reality*, but enough to destroy Hiroshima and 66,000 human lives with a single bomb. This is heinous, not something to be proud of.

Nuclear fission does not need to destroy. The first nuclear power plant went online in Britain in 1956. Now, there are 437 reactors in 31 countries providing 12 % of the world's electricity. That number is tapering off since the terrible Fukushima accident in Japan. Now, Germany is decommissioning reactors they have already built, and it seems unlikely the Japanese will build more. Handling the reactor waste as well as the radioactive products when something goes wrong has turned out to be a bigger problem than people thought. *Plus ça change, plus c'est la même chose.*

Fusion

Please take another look at the left hand side of the Fig. 1 curve. If I can *fuse* two H atoms on the left hand side all the way up to carbon, I'll get as much as 8 MeV per nucleon, eight times what we get from fission. Our Sun does that.

The most we've been able to do so far is fuse deuterium (H2) and tritium (H3) into helium (He4), producing 3.5 MeV per nucleon on average, for a total 17.6 MeV, including a 14.1 MeV neutron. That's pretty good. It is the basis of the hydrogen bomb, in which 0.7 % of the mass is converted into energy.

Thank goodness, that weapon has never been used in war (Fig. 3).

Safety and Utility

There's a major safety difference with nuclear fusion. There is no possibility of a catastrophic accident in a fusion reactor, unlike in fission reactors. There is no such thing as a fusion chain reaction. Fusion takes effort. The moment you stop applying that effort, the reaction stops. With fusion, safety is inherent. The reaction itself makes helium, which doesn't get radioactive. The reactor itself may be "hot," but

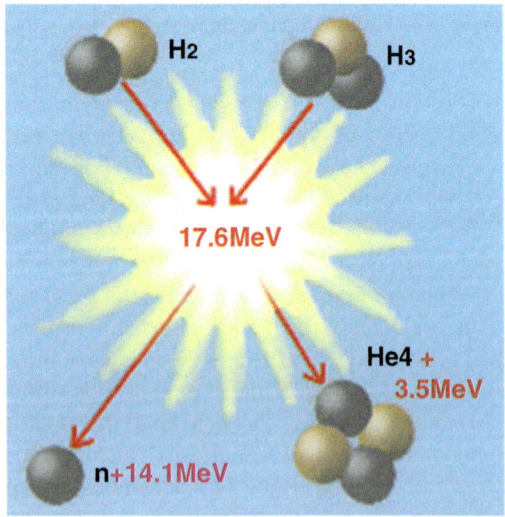

Fig. 3 Fusion (F. Wicke)

leave it alone and it'll cool a hundred times faster than do fission products. Also you can build the reactor out of stuff like carbon composite, that is strong but has a short half-life when it becomes radioactive. That promise is why I was working on "laser fusion" for many years.

By comparison, in Fukushima, when fission fuel rods lose their cooling, the reaction keeps on going for months or years, and can lead to melting, pooling, and boring a hole to hell.

A *controlled* fusion reactor is more useful than a fission reactor because there's enough of the heavy hydrogen isotope in seawater to power our planet for the next billion years, and you don't cause lung disease digging up the fuel.

Fusion Zoo

There are several ways to make isotopes of hydrogen fuse. There are other elements, such as He3, that could be fused with hydrogen. It's humbling that it's been so hard to do. All you have to do is bring the components together with an energy of 20 keV or so, not even a MeV, to get fusion to happen. So far, nobody has achieved *breakeven*, the level at which net energy is produced.

Inertial Fusion

You can do it fast, fusing the hydrogen before it can get out of the way. That's *inertial fusion*, and it's the reason for the National Ignition Facility (NIF) described in *Lasers*. NIF was built to do that with a few ns blast of laser light, using a tiny amount of hydrogen fuel. It's also the way the bomb works, with a much larger amount of fuel. So far, this approach has made no more than 27 kJ. At least it has made more fusion energy than went into the fuel. The hooker is that not much energy went into the fuel in the first place. After all, 420 MJ went into the capacitor banks that drive the giant NIF laser. But ... this is a *very nonlinear* process, meaning that a little more efficiency will make a humongous difference. Getting just 20 times more energy into the fuel capsule will generate 150 MJ of fusion energy in a single pulse, 5500 times more. Right now, the fuel core temperature is 3.5 keV, and you need more. For a *reactor*, you need a laser that can fire several times a second instead of several times a day. That would use laser diodes instead of flashlamps to pump the laser glass.

Magnetic Fusion

Magnetic fusion does fusion slowly, in seconds, with a powerful magnetic field to hold the plasma in place while it reacts. You can't hold a 400 million degree plasma with a container of solid material. That approach brings a flock of problems that inertial fusion avoids, but so far it has been more successful. The TFTR at Princeton

Fig. 4 Tokamak fusion test reactor, TFTR (Princeton Plasma Physics Laboratory, Creative Commons License via Wikimedia Commons)

(Fig. 4) has reached a temperature of 44 keV and produced 10 MJ of fusion energy in a one second burst. JET, the Joint European Torus has produced 16 MW for more than a second, giving 22 MJ. Amid all the hoopla, it's important to keep the numbers in mind. Still, they have not made more fusion energy than what went into the fuel from the neutral beam injectors.

Lots of other ways to get controlled fusion have been proposed. *None* have yet panned out.

Cold Fusion

Well ... that was a joke. In 1989, two chemists called Fleischmann and Pons announced that they had done fusion in a coffee cup, and lots of people wanted to believe that. After all, *what if* all that money to build NIF and JET had been wasted, and you could actually do it by sending a current through heavy water (water made from that second isotope of hydrogen and oxygen) between some palladium electrodes?! The announcement was *so* exciting! The oil crisis was a recent event. I was not immune. In the Japanese lab where I was working at the time, I spent 2 months trying to figure out how this could happen theoretically, assuming it was a fact. The copy of a mere abstract that we received had been faxed and refaxed so many times that it looked like a negative print of toothpaste splatter on a bathroom mirror.

That, by the way, is how *really interesting* science gets transmitted, compared to the months-long process of peer review, corrections, typesetting, and so on that greets normal papers! Makes you humble. Someone in our lab wanted to try it, but all the palladium in Japan at that moment had already been bought up! Even if it was BS.

Muon-Catalyzed Fusion

When they heard the Fleischmann-Pons news flash, people were receptive because they had heard about *muon-catalyzed fusion*. This was already predicted theoretically by Sakharov and Zel'dovitch in 1950. It turns out you can briefly make a Frankenstein heavy hydrogen molecule in which the two electrons are replaced by muons. The muon is a negative particle and it fits right in where the electrons were, except that it's 200 times heavier and makes an orbit 200 times closer to the nucleus. So, in a molecule, the two heavy hydrogen nuclei are 200 times closer to each other than they ought to be, and they sometimes fuse all by themselves at normal temperatures!

Wow! Let's go do it! Well, muons only live for 2 μs and you only find them in accelerator beams at Los Alamos and similar places. People have got 150 fusions out of each muon before it dies. But, this process is still a curiosity, not likely to ever be a power plant.

Ion Beam Fusion

Ion beams can be substituted for the laser beams in NIF to illuminate the fuel capsule, make X rays and implode the hydrogen target. The so-called "Z-pinch" at Sandia National Labs in Albuquerque has produced 3.7 billion degree temperatures and pressures of 100,000 atmospheres, launched flyer plates and so on, but it is difficult to find data on output fusion energy.

Electrostatic Fusion

The idea is to make a spherically convergent electric field between electrodes that will accelerate hydrogen ions and cause them to collide and fuse. It is similarly difficult to find data on output fusion energy for this technique.

Conclusion

People would *really* like to achieve controlled hydrogen fusion so they can make power for the next billion years or so. A lot has been spent on it. So far, except in the bomb, we are a very long way from doing it.

Part IV

Odds and Ends

Religion

However far back we look, people have worshiped a Creator or Higher Being of some sort. This was the solar deity Amun-Ra for the Egyptians (Fig. 1), Marduk for the Babylonians (Fig. 2), Baal for the Canaanites, Huitzliopochtli for the Aztecs, and so on. I think I would run fast if I saw Ra coming down the street!

Many people have believed in one God. The hairy Anglo Deity at the front of Chap. 4 is a caricature, but it's included because I so love the power of the artist's work. Gods are usually depicted as anthropomorphic.

Today, billions of people believe more or less exclusively in one of a few religions (Fig. 3).

If you read this far, you know that although I believe in God and pray, I am skeptical of idols and medicine men, and fall into that 15 % in Fig. 3 who are technically nonreligious, because I do not belong to a particular worship tradition.

If there were only one true way, there would only be one religion. Many people that I know claim that mankind needs religion, and I do not doubt that, considering the number of old friends I grew up with in northern California who have rejected a formal religion but are disturbingly religious about the things they will eat. I do not doubt that the religion you are immersed in and the culture or society in which you are immersed are strongly connected.

A second reason I don't doubt that mankind needs religion is that I also believe religion is the fiber of society. Religious teaching gives an idea of right and wrong behavior that puts fear of punishment, now or later, ostracism, etc. in place of personal gratification as the highest guidance for life. When that is missing, societies seem to become flaccid and shallow.

A third reason is that most religions promise life after death, solving in one stroke the ultimate mystery and contradiction involved in being brought to life without our choice, only to die. This is a powerful reason for the development of religions.

As a fact, millions have died even in modern times because people hate and fear others who adhere to a different religion, despite the commands to love, respect and

Fig. 1 Amun-Ra (Jeff Dahl Creative Commons License via Wikimedia Commons)

Fig. 2 Marduk (Rmashhadi Public Domain via Wikimedia Commons)

Conclusion

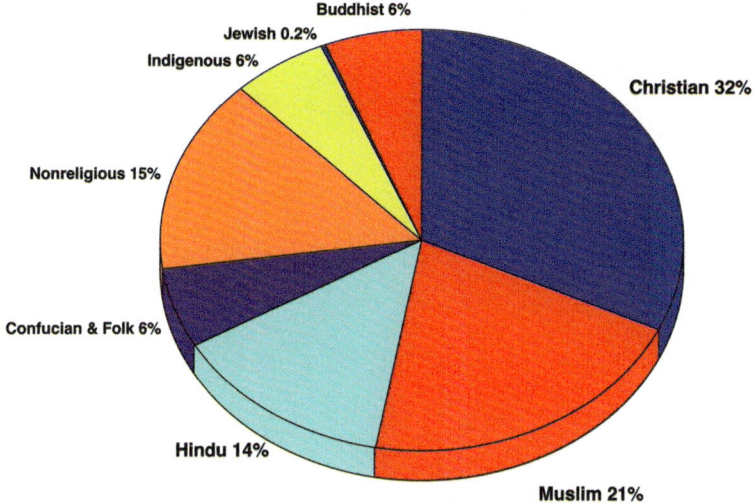

Fig. 3 Religions today (C. Phipps)

honor other humans in most religions. The thing I regret most is that humans, inflamed about a supposed affront to their own religion, have had so much trouble with the commandment "thou shalt not kill."

Priesthoods

The most significant priesthoods in the USA today, who claim power and right beyond that of ordinary mortals, would seem to be scientists and the leaders of the military-industrial complex rather than priests in the religious sense. Some of that is earned and useful, and some not. Many of these wield power well beyond what any ancient priest could muster.

The idea of democracy, control by the people through elected representatives, one of the key results of the Enlightenment, seems distant in many countries. Many do not even want it, and we should keep that in mind.

Conclusion

The most powerful forces in history are ideological, and have caused the death of millions.

Everything Else

This is the chapter for everything that is still fascinating but didn't fit elsewhere, and wasn't worth a whole chapter by itself: the Death Asteroid, where water came from and how you find exoplanets.

The Death Asteroid: The Really Big Space Debris Story

Did you watch the news a couple years ago about the meteor that blew up over Chelyabinsk in Russia? At 13,000 tons and 19 m in diameter, it arrived moving at nearly 20 km/s, like a lot of them do. Friction with the Earth's atmosphere made it explode while it was still 30 km above the Earth with the energy of a 500 kt bomb. Windows were blown out in a radius of several km, and more than a thousand people were injured. Warehouse roofs were collapsed (Fig. 1)—not by the meteor, but by the blast wave. People smelled sulfur all day.

About a century earlier, in 1908, a larger meteor than this one flattened forests in Tunguska, Siberia (Fig. 2), releasing the equivalent of 10 Mt. Fortunately, so far as we know, nobody was there. At an incoming velocity of 30 km/s, the object had ten times the energy density of dynamite! I mean, even high power rifle bullets only go 2 km/s and energy goes like velocity squared.

What is an asteroid, and where do they come from? Asteroids are minor planets flying through the solar system, usually bigger than 1 km. They can be ice, stone or iron. They drift into Earth's neighborhood from out beyond Jupiter. They can be pretty hard to see because of their small size and dark surface. Thanks to NASA's "Near Earth Object" observation program, scientists believe 93 % of those that threaten Earth have been discovered. Many are in orbits similar to Earth's and cross our orbit from time to time, and long observation makes their paths easy to predict.

How often do they hit Earth? The bigger they are, the less probable (Table 1). That's a good thing. Also, let's get *this* straight: the chance that you'll die from a lightning strike is much higher than that the Death Asteroid will get you. Still ….

Fig. 1 Asteroid explosion collapses Chelyabinsk warehouse (Posper A. Creative Commons License via Wikimedia Commons)

Fig. 2 Tunguska after its meteor strike. Fortunately, there weren't many people around. Thousands of km^2 were burned and flattened. An atmospheric shock wave from the blast circled the Earth twice, and dust from the impact obscured skies in London (Leonid Kulik, 1929. Public Domain via Wikimedia Commons)

Table 1 Asteroid impact energy and interval vs. diameter (ice 30 km/s, equivalent to stone at 10 km/s)

Diameter (m)	Kinetic energy (ice, 30 km/s, Mt)	Disaster interval (years)	Example
10	0.06	3	Chelyabinsk
100	70	1 k	Tunguska
1000	70 k	250 k	Meteor crater
10,000	100 M	60 M	Dinosaur extinction

End of the Dinosaurs

Now imagine a meteor 10 km in diameter hitting the Earth. Have you heard the story of how the dinosaurs died, almost overnight? Roasted or starved, depending on where they were. The scientist Luis Alvarez was wondering about that in 1980.

He was fascinated by the story from geologists that a very thin layer of the metal iridium could be found all over Earth at a certain depth called the K-T boundary which corresponded, geologically speaking, to a time 66 million years ago. This is the boundary between the Cretaceous and Tertiary periods and marks when the big dinosaurs all died. It would be the C-T boundary except that Germans named it.

What was the connection? He knew there was one, but it puzzled him for days.

Iridium is very rare on the Earth's surface, but asteroids have a lot of it. Alvarez suggested a truly giant asteroid hit the Earth at that time, completely changing life on Earth. And when he calculated the diameter of an asteroid that could contain that much iridium, the answer was ... 10 km, the size of Manhattan! At first, people thought he was nuts, but years of research have supported his theory, and even shown where the giant asteroid hit, in the ocean off the Yucatan peninsula in Mexico.

Radar shows a 180 km diameter crater there we now call the Chicxulub Crater, created by this asteroid impact, 65 million years ago.

You can't imagine how terrible the results were. The asteroid bored right through the ocean and ejected a giant plume of glowing lava into space that circled the Earth for several hours. Everything that wasn't inside a cave was exposed to a 1400° glowing oven as all this lava re-entered the atmosphere. At that time, the atmosphere was 35 % oxygen, causing combustion more intense than anything you've seen. Then, because the ocean bottom where it hit was gypsum (a form of calcium sulfate) there was acid rain and sulfuric acid aerosol, and later a worldwide dustcloud which darkened the sky for 2 years, completely terminating the growth of plants that had not already been burned. This included phytoplankton in the ocean. This was the "nuclear winter" from hell. After the dustcloud settled, what was left was less oxygen and a lot of CO_2, creating a classic greenhouse effect.

How did *anything* survive? Little creatures in caves that didn't need a lot of food, buried seeds, things that hibernate in the ground like cicadas, fish burrowed into the seabed—these survived. Gradually the buffering effect of the Earth's rocks and soil neutralized the acid, the dust settled or rained out and life returned to normal, but without the big creatures.

It could happen again. Figure 3 shows the orbits of the ones we do know about. But, as with everything, it's the Black Swans, the ones you *didn't expect* that you worry about. You worry if you discover this thing approaching Earth for the first time from the direction of the Sun, where it's hard to see. At 30 km/s, you might have just 100 days to respond.

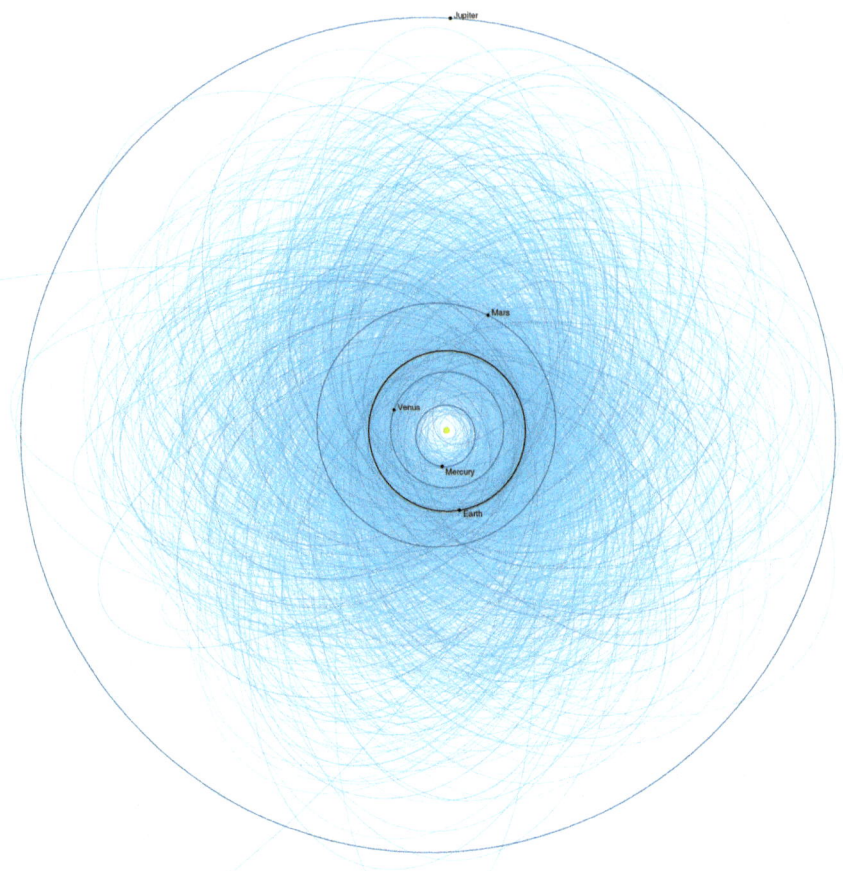

Fig. 3 Potentially hazardous asteroid orbits. There are a lot of them! (Jet Propulsion Laboratory NASA, public domain via Wikimedia Commons)

Solutions

What do you do? Scientists have been considering two solutions. What you favor depends a lot on what you know about. One is to fly up and deflect it with a bomb. But you'd better be careful. If you get too close and fragment the thing, you've made a cloud of fragments, and that would be worse—a million balls of fire instead of one big one.

This thing is going 30 km/s, and it's hard for rockets to match that speed. But let's say you do. The best you could do is whiz past it at a relative velocity of 60 km/s as you meet it halfway with 50 days remaining. You'll have exactly 1/6 of a second to do something and you'd better do it right because you won't have a second chance.

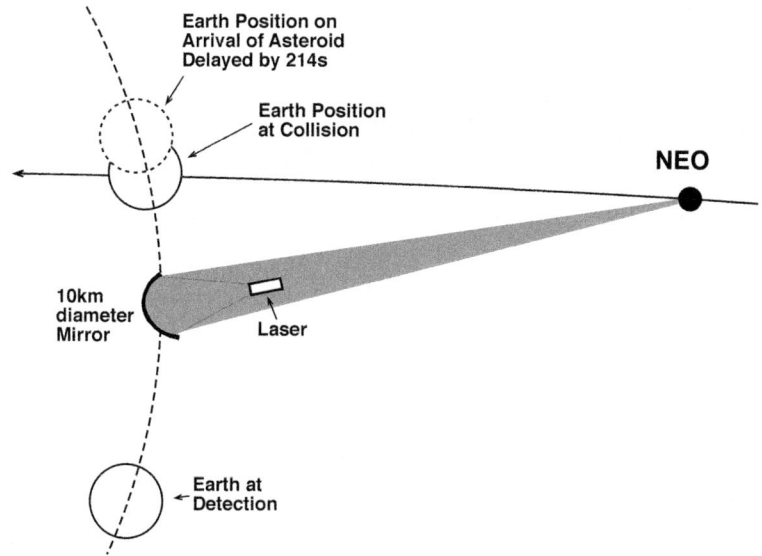

Fig. 4 Planetary defense with a laser (C Phipps)

The second idea is to deflect it with a giant laser. If you find an asteroid destined to hit the Earth dead center, just push it gently for 6 months. You won't crack it up. Your beam gets to it at the speed of light, not the speed of a rocket (Fig. 4). You slow the asteroid just a bit, so it arrives about 3 min late and the Earth has already moved out of the way! You can make the mirror out of aluminized mylar stretched in a spinning frame. The laser would cost ten billion dollars, not a bad price for saving Earth. In Fig. 4 I'm just showing the laser station hanging out in Earth's orbit for simplicity's sake. Actually, it should be at one of the Lagrange points L4 or L5 considerably ahead or behind the Earth in its orbit and, of course moving with the Earth. These are places where its position could be stable, and it would still have a good view of the cosmos.

Asteroids have hit us many, many times, and a lot more often, thankfully, when the Earth was younger than now. That makes sense because the Earth was formed by sweeping up smaller junk and the more you sweep, the fewer there are.

But we need to be *thankful* to asteroids. Why? …

Where Did Water Come From?

Ah, we've always had oceans on Earth, right? Nope. Back when it was a glowing orange rock a few billion years back, certainly not! So where did water come from?

Well, we don't know! Did you know that?

The most likely possibility is from the steady rain of those scary asteroids, because it turns out a lot of them are iceballs.

Another really weird way we could have gotten water is that Earth started out with *a lot* of hydrogen sulfide and CO_2, which bacteria combined into water, sulfur and ... *formaldehyde*. A sea of formaldehyde? Eeesh!

Anyway, we now have oceans of blue water, miles deep in some places, and we should be thankful for it and not just dump plastic bags in it. We won't get any more.

Finding Exoplanets

If we are looking for oncoming asteroids, small things very far away, we need big telescopes. If we have big telescopes, we can also look for ... exoplanets! *What?* An exoplanet is a planet around a star other than our Sun.

Planets and Stars

Some of you are saying "I didn't take astronomy yet. What's a planet?" A planet is a smaller, colder object like Earth looping in an orbit around a star. Our Sun is not the only star that has planets going around it.

What Is Special About Earth?

Throughout this book I've been saying we're nothing special here, not the center of anything, surely. But, up until this year, we thought we were special in one way: our big Moon stabilizing our orbit around the sun.

Next full Moon you see, thank it for making the Sun rise in the East instead of over the poles. *What*? That's a bit of an exaggeration, but our Moon going around us is a stabilizer of our axis tilt. Axis tilt is what makes summer different from winter in northern and southern temperate zones. And, *stable* axis tilt is why the old Earth has had repeatable winters and summers for many billion years while life developed—jungles in the tropics, polar bears in the arctic.

Other planets have moons, but very few as big as our Moon is relative to the Earth, one quarter its diameter. That's because it was not made by junk collecting in a pile but by a Mars-size planet that crashed into us early on, making a blast of molten stuff that congealed into the Moon. Why is that difference important? Because computer simulations show that the tilt of our axis might change randomly by 20° in half a billion years without the Moon. Right now, it's 23° and has stayed that way for a very long time.

Even though where it points changes slowly over 26,000 years. And that, dear friend, is why sight-holes in the Pyramids don't point where they used to when they were built, and why the North Star will be Vega in 13,000 AD. We can thank the presence of our Moon for that predictability. But the Moon is not why our orbit is a stable circle around the sun, which is what we need for life to develop. *Jupiter is*!

Contrary to what we though just 5 years ago, this means: *the odds for life on a Moonless Earth-sized planet are good.* Even in that way, we are not special.

Remember when I asked you to lay out the solar system to scale on a 10-m-long piece of paper in *Metric System*? My point there was to help you understand how insignificant our little speck is in the Cosmos. And yet, Jupiter stabilizes our orbit! Almost makes you believe in astrology.

Stars

What's a star? This is a hydrogen bomb that's going off all the time, so far the only successful fusion reactor turning hydrogen into helium into lithium into carbon into calcium and so on, over billions of years. On the surface, our sun looks like 5700°K and glows yellow-white, hardly a nuclear reactor, but in the center it's tens of millions of degrees and hundreds of billions of atmospheres pressure, quite enough to fuse elements. It's making several billion kg of new stuff every second. It's an average star, so it will live a long time.

Planets and Exoplanets

The reason people say "we are stardust" is that these smaller colder objects like our Earth, full of calcium carbonate, iron oxide and so on wouldn't be here unless generations of other, older stars had not lived their lives and exploded in a supernova, spreading the heavy elements they made around the universe! Our own sun doesn't expel what it makes until it explodes! These collect and collect in bigger and bigger piles held together by their own increasing gravity, then *sometimes* end up orbiting another star, which is where we are today. But without those elements, we ourselves wouldn't exist. You can't make people out of hydrogen, so far as we know!

We know exoplanets must be there, because there can't be anything really unique about our Solar System, even though we've wanted to think so for thousands of years. You can't *really* see those exoplanets. At least not yet. To see planets around even nearby stars, you'd have to build a several hundred meter diameter mirror. That's because the very closest star is still 4 light years away, and the nearest one that has a planet is 12 light years distant. *That's 750 times as far as the Earth is from the Sun!* People do think about building such huge mirrors (Fig. 4!) but so far they're theoretical.

Figure 5 shows that we have now found more than 1000 of them, and by the time you read this there will be a lot more. And now thanks to the new knowledge about Jupiter's influence, lots of these planets can be assumed to have life! Isn't that *amazing*?

The rate of discovery is growing fast. So, how did we do it? There are at least five ways, all of them ingenious.

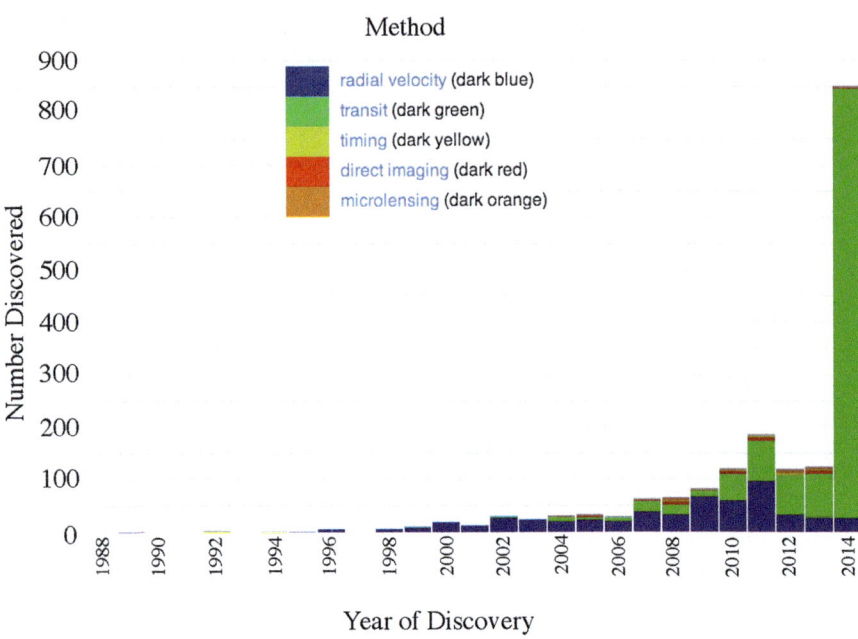

Fig. 5 Exoplanets discovered vs. time (Aldaron [Public domain], via Wikimedia Commons, axis labels modified so you can read them by C. Phipps)

The Transit Method

If you're lucky, you find a planet with an orbit that happens to be edge-on viewed from Earth. You find a planet by finding unexpected changes in the brightness of the star that could only be explained by the planet momentarily passing in front of its sun. If it's a small planet in front of a big star, the brightness change might be only 1 %, but people can do that.

Radial Velocity

This is just Doppler spectroscopy. You see the frequency of a radiated spectral line from *the star* go up and down when it's being dragged toward or away from us by its planet. Here, it's not absolutely necessary that the planet's orbit is oriented edge-on toward the Earth, but it helps. If we're looking perpendicular to the orbit's plane, we don't see anything because there's no velocity change relative to us. If there are many planets, it gets more complicated.

Timing

If the star is a pulsar, you can see changes in the timing of its pulses because a planet is going around it. The only trouble is, this is a rare case.

Direct Imaging

We don't really mean this. Sometimes a planet is far enough away from its star that you can see a blob of light moving around the star's blob. This is amazing, but still a long way from seeing the planet, like we see Jupiter.

Microlensing

See Fig. 15 of *Modern Science*. Gravitational lensing is something we talked about when we talked about general relativity. Now imagine a really unusual situation in which two stars are aligned so that the one nearest us is lensing one behind it, *and* the front one has a planet. If you're careful, and fast because the alignment might not last more than a few weeks, you might see some funny changes in the image that come from the planet making the front one wobble.

Among these, about 30 seem to be *habitable* by the standards we have for life. About half of these are Earth-size, so you wouldn't have to be a monster to move around, because some are very big compared to Earth and have a lot of gravity. Sadly, the average distance is 650 light years. Many have two suns! Think about that! They orbit a binary star system in which two suns orbit each other! Wouldn't that be beautiful? About four are within 20 light years, so there is some hope of at least communication if not visits.

It's been more than 20 years that we have been emitting light and radio waves that someone else could see. So, it would appear that we are alone in the immediate neighborhood. Why has no-one contacted us? Planet tau Cet e is only 12 light years away. One possibility is that, after taking billions of years to evolve and becoming technologically sophisticated, civilizations wipe themselves out, or become fat and happy and *are* wiped out, pretty quickly. So the *probability* that a planet we find right now can hear us and respond might be 50 years/4 billion years. See the movie *Forbidden Planet*.

Another one is the fear of contagion, on *their* part. See *The Day the Earth Stood Still*.

Conclusion

Big asteroids are out there, and have played a major role in the history of our planet, both for good and ill if you'd like to see a live dinosaur. The ones that are like a hydrogen bomb seem to come every 50 years or so. But our planet is big! More often than not, they lay down a forest but nobody is killed. If you see it coming and you have enough time, you could bomb it or deflect it with a big laser. Asteroids probably brought water to the Earth. You need a big telescope to see them coming. And if you have one of those, that's at least one way of seeing other worlds around other suns. So far, we've found more than 1000 of them. The biggest asteroid of all crashed into us very early and made the Moon, which is a good thing.

The manufacturer's authorised representative in the EU is Springer Nature Customer Service Centre GmbH, Europaplatz 3, 69115 Heidelberg, Germany. If you have any concerns regarding our products, please contact ProductSafety@springernature.com

Printed and bound by CPI Group (UK) Ltd, Croydon, CR0 4YY

25/03/2026

02078177-0006